PRACTICAL WORK IN ELEMENTARY ASTRONOMY

M. G. J. MINNAERT

PRACTICAL WORK
IN ELEMENTARY
ASTRONOMY

SPRINGER-VERLAG NEW YORK INC. / NEW YORK

D. REIDEL PUBLISHING COMPANY / DORDRECHT-HOLLAND

SOLE DISTRIBUTOR FOR NORTH AND SOUTH AMERICA

SPRINGER-VERLAG NEW YORK INC. / NEW YORK

Printed in The Netherlands by D. Reidel, Dordrecht

TABLE OF CONTENTS

PLANETS AND SATELLITES

B.   THE STARS

THE SUN

THE STARS

ACKNOWLEDGMENTS

Remembering our lively evenings of practical work, I cannot but express my deep appreciation to the instructors and assistants of Utrecht Observatory, who devoted so much of their energy and enthusiasm to the development of this teaching. My special thanks are due to Dr. H. Hubenet, who organized and directed the course during the later years, and who has been extremely helpful by supplying experienced advice on so many points.

TO THE INSTRUCTOR

Astronomy is a science of nature. It is based on observation, and it is to the results of the observations that theory and calculation are applied. Our practical work, therefore, will have to show the concrete reality of the celestial objects, studied during the lectures: as far as possible this practical work should run parallel with the theoretical courses.

Practical work in General Astronomy can be organized along very different lines. We shall mainly describe the course which has been developed at Utrecht during a period of some 25 years. It is intended for freshmen; future mathematicians, physicists and astronomers, who from the very start should be confronted with the sky before they are asked to look at the blackboard! They are expected to know some trigonometry, the elements of calculus and physics; but the astronomical problem will always be put central.

The students are formed into groups of about 25, working in pairs; each group on a fixed evening of the week.

By letting all the students work simultaneously, a rather close synchronization with the course on General Astronomy becomes possible. The introductory explanations can be given collectively; a comparison between the results allows an estimate of the accidental errors and introduces stimulating competition. On the other hand, care should be taken to leave sufficient time for free individual initiative; the paragraph numbers, put within parentheses and the literature references are especially intended for such purposes. After all, there is no obligation to finish each exercise within just one evening.

The principle of simultaneous work has the consequence that each instrument must be produced in quantity: one for each pair of students. Consequently, only simple instruments can be made available, such as will be described in the beginning of this volume.

Whenever the sky is clear, observations are made from a terrace in the open air. Directly afterwards the results are used as a basis for simple calculations. Observations through a medium telescope, placed inside a dome, are made in between the other parts of the programme; the students are called two by two, and have to interrupt their work only for a few minutes.

When cloudy, photographic records or other documents are studied, instruments tested, or practical calculations carried out. These are no scholastic exercises: the professional astronomer also devotes a great part of his time to such laboratory work.

Two programmes are prepared for each evening, one of each kind, and a choice is made one hour ahead. Each exercise in this book describes such a programme, mostly planned in such a manner, that it can be carried out in about 3 hours.

In general it will not be possible to carry through the whole programme of this book;

a choice should be made according to weather conditions, instruments available, predilection of the instructor, and number of hours available. The choice will also be different for those students who specialize in astronomy and for those primarily interested in mathematics or physics.

It would be desirable to have the students working in the library, surrounded by books and periodicals, consulting any sources they wish and getting the material first-hand from the original publications. In our experience this has proved possible as long as the number of students was small, but even then the material became partly spoiled. With greater numbers of students it was unavoidable that a great part of the reference material had to be reproduced by photocopy; these photographs were then protected by a thin cellophane sheet.

When starting this practical work in astronomy, we were inspired by the wonderful early work of H. T. Stetson, R. K. Marshall, O. L. Dustheimer and other American astronomers, from which the basic ideas for some of our more elementary exercises are borrowed. These remain necessary as long as astronomy is not generally taught in secondary schools. A good instructor will nearly always be able to make them 'scientific'; the literature references may be of use for this purpose. On the other hand, European university education does not correspond to the American college and requires a somewhat more thorough treatment for more specialized students. For those less elementary exercises no examples seemed available and we had to find our own style. Avoiding therefore a course of only theoretical little problems, we have inserted direct observations, the study of astronomical photographs, and the use of simple instruments, as often as allowed by the climate and the available instrumental means. Our practical exercises have developed all the time in the course of the years, according to experience gathered in working with the students and to ideas suggested by the instructors. A few instruments, especially constructed for our practical course, were successively improved in constant consultation with our workshop. Descriptions will be found on pp. XV–XXIII.

Let me express the wish that practical work in elementary astronomy will soon be generally introduced in university teaching and that our experiences, here described, will prove of some use for this purpose. We have found it most enjoyable to build up such a course and we are sure that others will find the same satisfaction in this creative work.

BOOKS ON PRACTICAL WORK IN ASTRONOMY
(Mainly for the elementary exercises).

DAGAJEV, M. M.: 1963, *Laboratorny Praktikum po Kursu Obshtshey Astronomii*, Moscow.
Deutsches Pädagogisches Zentralinstitut: 1962, *Praktische Schüler-beobachtungen für den Astronomie-unterricht*, Berlin.
JASCHEK, C.: 1968, *Exercises in General Astrophysics*, in press (for more advanced students).
SHAW, R. W. and BOOTHROYD, S. L.: 1958, *Manual of Astronomy*, Brown Co. Publishers, Dubuque.

TO THE STUDENT

A student, following a course in astronomy, expects that now at last he will see with his own eyes the wonders suggested by the scintillating stars in the depth of the night sky. This expectation will be fulfilled, though some work and effort will be necessary. To observe through a telescope requires *exercise*. The study of photographs is a study of *symbols*, of which the real meaning has to be discovered by reflection. During the work one should try to realize how immense, how harmonious is the structure of the Universe which we are trying to explore. You may be sure that the professional astronomer has the same feeling of wonder and awe. Seldom will he speak about this, but it inspires him all the time in his work.

For a series of selected topics our practical work will demonstrate the methods which are used in the investigation of the Universe, not getting down to technicalities, but concentrating on the principles. It should convey to the student some idea of the work of the astronomer in his professional practice. It is not primarily intended to teach technical skill, but the student should learn by practice the style of scientific investigation and the methodology of research work.

We shall work with very simple, home-made instruments. This is necessary, since many of them are needed; but it has also the advantage that the essentials of the method become more apparent. However, the student should understand that he must always endeavour *to reach the highest precision attainable* with a given instrument even when this instrument is primitive. The following implements are constantly required:

compasses, big protractor, slide rule;

a note-book without lines; a copybook; sheets of rectangular co-ordinate paper (by preference red).

The students work two by two. Usually one of the partners observes, while the other records the figures; after each series the roles are reversed. These records should be made *orderly* and *methodically* in the note-book, and should in no case be corrected later. Then each partner writes a short report in his copybook, containing all the observations mainly *in tabular form*. (Many examples will be given; you are free to arrange your results in other ways.) To make a good scientific report requires special skill and experience, which should be acquired early. Bear in mind that after three months your report should still be readable and understandable!

Not all students are equally quick. The tasks, therefore, are arranged in such a way, that the first paragraphs are the most essential ones, and that following paragraphs, indicated by numbers between parentheses, can be chosen as interesting complements for those who have finished the main programme. Any personal initiative will be encouraged!

Graphs and calculations are made by each of the partners independently, the results being compared after each step, which gives a good check: professional computers often do the same. It is preferable to have the reports made directly after the observations, while these operations are still fresh in the memory.

Drawings of observed objects should be made by each partner on a sheet of the note-book and then glued into the copybook (glue at the topcorners only!). No artistic ability is required to make such a drawing; it helps you to realize what you are seeing and to express your conceptions by simple sketches.

This is not school, it is scientific research, albeit at an elementary level. We work in order to *understand* things. The assistants will not rate your merits, they will just try to explain and to guide. You can use books and lecture-notes, you can consult the observatory library, you are invited to help each other and to discuss questions that might not be clear, but conversation, not related to our work, would be disturbing.

In our Instructions:
Observations on the *sky* and operations in the *laboratory* are distinguished by the characters *S* and *L* after the paragraph numbers. Numbers between parentheses refer to paragraphs, not essential for the main problem of the exercise.

General References

ALLEN, C. W.: 1963, *Astrophysical Quantities*, London.
BRANDT, J. C. and HODGE, P. W.: 1964, *Solar System Astrophysics*. McGraw-Hill, New York.
CHAUVENET, W.: *A Manual of Spherical and Practical Astronomy* (several editions; reprinted in the Dover Publications, 1960).
DANJON, A.: 1952–53, *Astronomie générale*, Paris.
DANJON, A. et COUDER, A.: 1935, *Lunettes et télescopes*, Paris.
KUIPER, G. P. and MIDDLEHURST, B.: 1953–1963, *The Solar System* (Vol. II–V), Chicago.
KUIPER, G. P. and MIDDLEHURST, B. M.: 1960, *Stars and Stellar Systems* (Vol. I–VIII).
UNSÖLD, A.: 1955, *Physik der Sternatmosphären*, Berlin.

OBSERVATIONS FOR AMATEURS

ROTH, G. D.: 1960, *Handbuch für Sternfreunde*, Berlin.
SIDGWICK, J. B.: 1956, *Amateur Astronomer's Handbook*, London.
A number of copies to be available of:
Norton's Star Atlas (Gall and Inglis, London, many editions).
The Astronomical Ephemeris; in many cases copies of former years may be used.

TECHNICAL NOTES CONCERNING THE PRACTICAL WORK
IN ASTRONOMY

For our practical work we need:

    (a) an observing terrace;
    (b) an astronomical telescope, mounted in a dome;
    (c) a 'laboratory', where indoor-work is carried out.

It is of great advantage if these three units are in *direct* proximity to each other.

(a) It is not easy to find a convenient *terrace* for students' observations. A free sky is of course desirable, a free view towards the South is especially important. On the other hand, it is a considerable advantage if the observing terrace is surrounded by low walls and by trees (at some distance), as protection against the wind and for screening off city-lights.

On the terrace we have a series of small pillars, one for each pair of students (Figure 1). They each have their serial number; on each of them a small experimental telescope with the corresponding number can be placed in a fixed position. The pillars are hollow and open from behind. Inside there are weatherproof plug points for alternating electric current 24 V.

Light metal chairs should be provided for exercises in which drawings have to be made.

Recently a series of small mountings have been constructed, to which a camera for celestial photography may be clamped; each may be rotated by a synchronous electromotor with the normal diurnal speed. We have no experience as yet with this installation.

(b) We use a *refractor* with an objective of 25 cm and a focal length of 3.50 m. It has a fixed connection to the gears; and a quartz-clock, giving sidereal time, is built in, so that declination and right ascension may be read directly from dials. The technique of working with graduated circles and of transforming hour angle into right ascension is learned by the students when they work with their small experimental telescopes; consequently it may be dispensed with when objects are demonstrated with a real telescope.

(c) The *laboratory* should be equipped with long tables, where atlases etc. can be laid out. Books and tables should be available, not only the elementary textbooks, but also some of the professional reference works. If possible some series of the important astronomical journals should be included.

The tables should have electrical contact-boxes for 24 V for the use of photo-meters or other instruments and 220 V for local illumination. Not too far from the terrace there should be two astronomical clocks, one giving Universal Time, the other Local Sidereal Time.

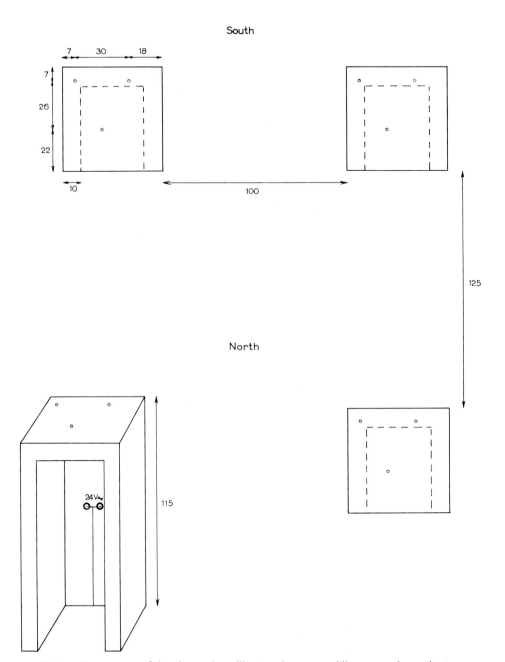

Fig. 1.   Arrangement of the observation pillars on the terrace. All measures in centimetres.

Some Students' Instruments, constructed in the Observatory workshop

1. *Experimental telescope* (Figures 2 and 3). The objective is a small achromatic lens, diameter 40 mm, $f = 500$ mm. The instrument is equatorially mounted. The polar axis is adjusted in altitude by one of the screws of the tripod. The adjustment in azimuth is obtained by means of a joint between the main vertical axis and the polar axis; this is clamped in such a way that the adjustment, once attained cannot be disturbed by the students. There are two circles, graduated with white lines on a black ground; the declination is graduated in whole degrees, the hour-angle in divisions of 6^m.

The ocular is of the Ramsden type, $f = 25$ mm. For some observations a stronger eye-piece is useful ($f = 9$ mm).

In the focal plane there is a fixed diaphragm of 15 mm and a reticule having two mutually perpendicular lines, oriented along the parallel and along the hour-circle (Crosswires were found to be too vulnerable). A simple dew-cap is found very useful.

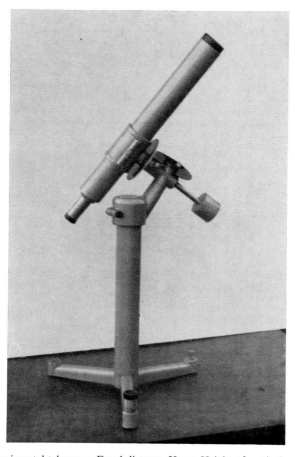

Fig. 2.   Experimental telescope. Focal distance 50 cm. Height of vertical column 45 cm.

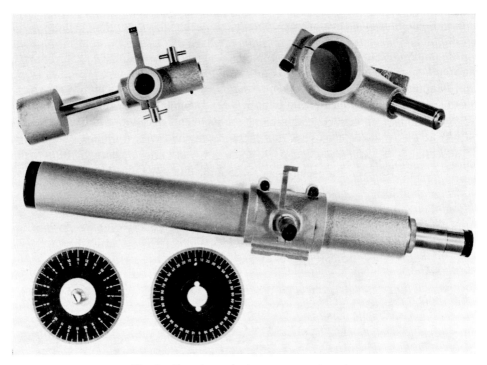

Fig. 3.   Experimental telescope; separate parts.

On top of the telescope tube a small level is mounted. Of course the declination axis should be accurately perpendicular to the polar axis and to the telescope axis, since errors in this respect cannot be corrected later. The zero-point of the graduated circles may be easily adjusted and fixed by an ordinary screw.

The three legs of the tripod are provided with blunt conical pens; one of them is a screw for the adjustment in elevation, it is fixed by a nut. The position of the three legs on their support is determined by the classic system: a hole, a groove, a plane – which removes any degree of freedom.

The position of the level has been carefully regulated. The orientation of the polar axis, the zero point of the graduated circles have been adjusted in first approximation and are checked in exercise A9.

2. *Microphotometer* (Figure 4. To be used in the exercises A31, B3, B5, B18). A small incandescent lamp, burning on 12 or 24 V~, is imaged by a system of 3 ordinary spectacle glasses on a small hole in the object table. The rays passing through this hole illuminate a photovoltaic cell, the current of which is recorded by a 100 micro-ampere meter, mounted on the foot of the instrument. By slight shifts of the lowest lens, the maximum concentration of the rays on the aperture may be obtained. In that case the microampere-meter reaches nearly its full deflection: the sensitivity of the instrument has been so adjusted.

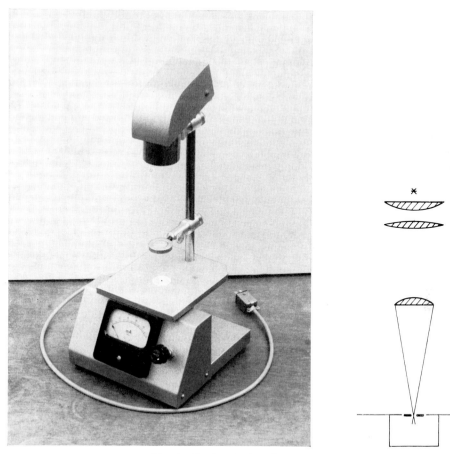

Fig. 4. Students' microphotometer.

[By using the modern solar cells it becomes possible to reduce considerably the brightness of the incandescent lamp, which is an appreciable advantage.

Another possibility is the use of cadmiumsulphide photo-resistors. The source of the electric current has then to feed the lamp as well as the photo-resistor.]

In order to avoid fluctuations of the tension, all students have to connect or to disconnect their photometers at the same time. It will be of course an advantage if, for each pair, the current is first rectified, then stabilized (stabilization to 1% will do).

The photographic plate to be investigated should be kept in closest contact with the object-table, the gelatine side should therefore be the downward side. In order to avoid scratches, a piece of smooth coordinate paper is put on the object table, leaving only the centre free. Edges of the metal plate should be rounded off and kept smooth.

3. A few pairs of *binoculars* are used as a transition between observations by eye and observations through the experimental telescope. Because of the wide field many

objects are even more impressive than when seen through a telescope. It is, however, necessary to keep the binoculars almost motionless by leaning against a wall or a door-post.

4. *Sextant* (Cf. exercises A13, A15, A20). Used marine sextants may be often acquired at a moderate price. It is our most precise instrument for the measurement of angles and it does not require a fixed stand. However, one has seldom a free horizon in view of altitude measurements. Moreover, inexperienced students often have some difficulty in getting the right field of view.

5. *Cross-staff* (Cf. exercise A20; Figure 5). This is an imitation of an old instrument, applied in elementary practical courses at Wellesley College, especially for the

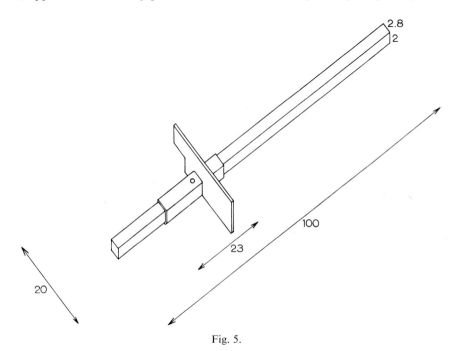

Fig. 5.

moon's orbit; later Marshall and Shaw-Boothroyd used it for the same purpose. Its use is described in exercise A20.

On the lower side of the staff two scales are found, corresponding respectively to the long and to the short side of the cross. Once the right adjustment is obtained, a screw is gently tightened in order to fix the position and the reading is made.

In establishing the scale, the distance between the eye and the extremity of the bar has been taken into account. The scale divisions correspond to steps of 1°, tenths are easily estimated. The scale is protected by a glass cover.

6. *Altimeter* (Figure 6. To be used in exercises A7, A14). In a triangular block of concrete a vertical iron tube *A* has been inserted, 63 cm high, 4.3 cm wide. A second, slightly narrower tube *B* (3.3 cm) inside the first one slides up and down and may

be fixed by a hand-screw. Tube *B* has near its end a screw, serving as an axis for the wooden measuring tablet (Figure 6).

This tablet has the size of 48×24 cm; it carries two dioptres *M*, *N*, a light pendulum suspended in *A*, and a scale *BD*, graduated in cm and mm. The distance between *A* and *BD* is taken equal to 20 cm, in order to simplify the computations. We direct the tablet so that the object is seen in the dioptres, we clamp the winged nut and read the intersection *C* of the vertical thread with the scale. – When not in use, the pendulum weight is inserted between two springs.

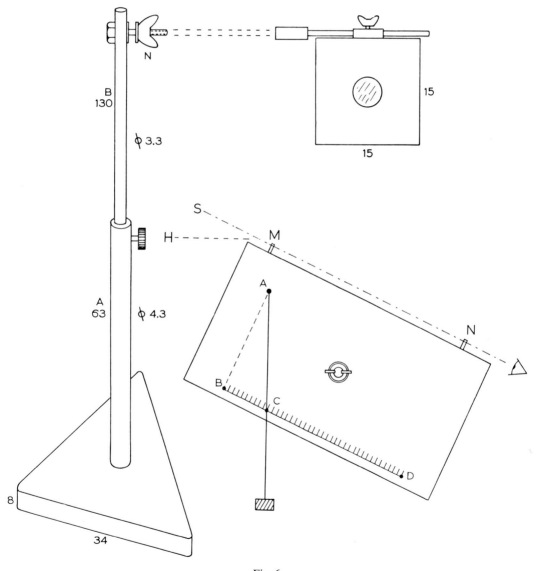

Fig. 6.

To the same axis at the upper end of tube *B* may be screwed a simple adjustable lensholder for exercise B1.

For exercise B2 the inner tube is removed and replaced by an articulated head, carrying our simple pyrheliometer.

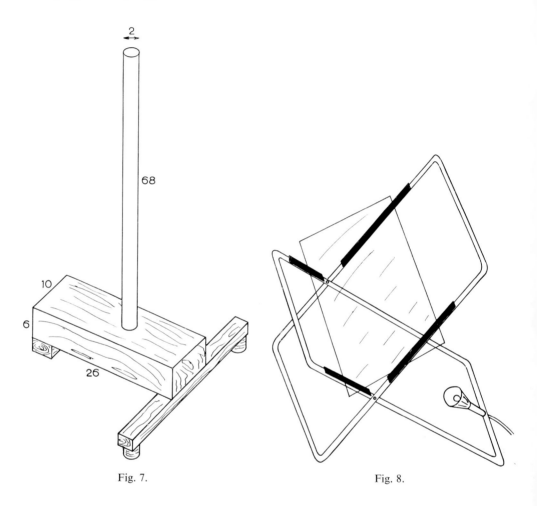

Fig. 7.   Fig. 8.

7. *Light wooden stands*, carrying a vertical iron tube, are used in order to clamp screens or diaphragms (Figure 7).

8. A *pulpit* (Figure 8), made of aluminium strips, is covered with plastic tube along part of the strips. On this pulpit a plate of milk glass may be put under some inclination, serving as a bright background for spectrograms or other photographic plates. The milkglass is illuminated from below by an incandescent lamp *L*, carried by a movable arm.

9. *Lamp*. For illumination of note-books or instruments with graduated circles and cross-wires, special lamps have been constructed (Figure 9). – They have a switch

S and can be damped with a resistor *R*. The light is emitted sideways from a hole *H*; one can choose between red or white. By means of a simple stand the lamp may be put in an inclined position, giving a faint light on the Star Atlas. On both ends the cylinder is protected by rubber rings.

Before electric current (24 V $\sim$) was available, we used flashlights, but then a regular and annoying check on the batteries was necessary. Dimmed light was obtained by using a lamp for which the battery gave too small a tension, or by inserting sheets of paper.

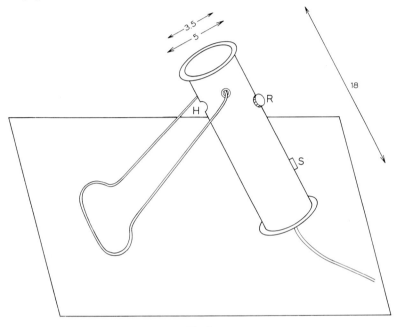

Fig. 9.

10. *Template* for plotting star positions (Cf. exercises A1, A2). A strip of copper, 4 cm × 1 cm, thick 0.12 mm. Holes are pierced with diameters of 2.2; 1.5; 1.2; 0.9; 0.5 mm.

11. *Microscale* for measuring the diameter of photographic star images. This is a photographic reproduction on glass of Figure 10, the enlargement being so chosen that the scale becomes a millimetre division.

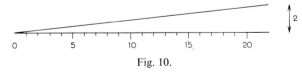

Fig. 10.

12. *Magnifier*. A plane-convex lens, mounted in a simple brass ring with a small handle. Students should be instructed to turn the convex side towards the eye and always to keep the eye quite close to the lens.

A. THE PLANETARY SYSTEM

SPACE AND TIME, INSTRUMENTS

A1. THE STARS AROUND THE NORTH POLE

The Problem

To become familiar with a few constellations around the North Pole. To draw a map of this part of the sky, making use of the tabulated star-coordinates.

Just as a city on earth is determined by its latitude and longitude, so is a star defined by its *declination* δ and *right ascension* α. Remember that r.a. is measured along the celestial equator, always in the sense W–S–E, starting from a conventional origin (Aries); the circle of $360°$ corresponds to 24^h of r.a.

Procedure

1S. *(If the sky is clear).*
From the terrace we observe Ursa Major and the Pole Star. The instructor indicates them by means of a strong flash light; the students standing near him can easily identify the stars towards which he is pointing.

Starting from the Pole Star, we notice in succession Cassiopeia, Perseus, Auriga, Ursa Major, Ursa Minor, Bootes, Cygnus. Notice the bright stars Capella ($=\alpha$ Aur), Arcturus ($=\alpha$ Boo), Deneb ($=\alpha$ Cyg).

Make a frequent use of 'alignments', by which the relative position of stars may be found.

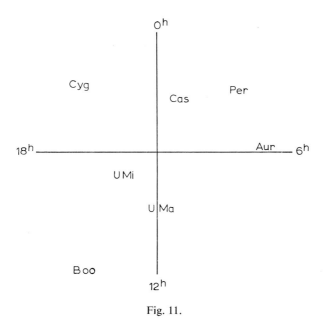

Fig. 11.

Now take a simple star map*, which you illuminate with dimmed light; find the observed constellations; compare with the sky.

Picture in your imagination the course of the celestial equator on the sky. Where are W, S, E, N? In what sense is the r.a. counted?

We now continue in the laboratory.

2L. Take a sheet of polar coordinate paper. The parallels will be numbered 80°, 70°, 60°, 50°, 40°, 30°; adapt the scale in such a way, that 2 cm = 10°.

3L. The paper should be turned till the longest side is vertical before you. The radius which is pointing upward will be taken to correspond to r.a. 0°; the other radii (representing the meridians) will carry their respective r.a. numbers; the finest subdivisions will probably be found to correspond with 8^m. Take care to number the meridians in the right sense! Remember that our maps have to represent the stellar sphere as it is seen *from inside*. Let the instructor check that your numeration is correct (otherwise your work would be useless).

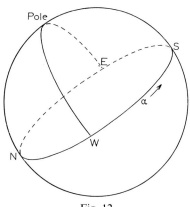

Fig. 12.

4L. We need the coordinates of the following stars:

Ursa Major (UMa) $\alpha - \beta - \gamma - \delta - \varepsilon - \zeta - \eta$
Ursa Minor (UMi) $\alpha - \delta - \varepsilon - \zeta - \eta - \gamma - \beta$ (α UMi = Pole Star.)
Cassiopeia (Cas) $\beta - \alpha - \gamma - \delta - \varepsilon$

These coordinates are found as follows:

(a) look up the star in *Norton's Star Atlas* and read provisional coordinates (precision $\Delta\delta = 1°$; $\Delta\alpha = 10^m$);

(b) find the star in the *Astronomical Ephemeris*, table 'Mean places of Stars', in which the stars are listed according to r.a.; read more precise coordinates ($\Delta\delta = 1'$; $\Delta\alpha = 1^m$).

This is a standard procedure!

Note at the same time the brightness of the stars, defined by their magnitude m.

For the constellations Ursa Minor and Cassiopeia, see Table I at the end of this exercise.

* Suitable for first orientation is the survey map on the front cover of Norton's Atlas.

5L. Plot the stars in your system of polar coordinates. You assign to each of them a dot, of which the size corresponds roughly to the brightness. Remember that *bright* stars have *small* magnitudes! A thin brass strip with holes of increasing sizes may be used as a template. Place the Greek letters, according to Bayer, and connect the dots by a thin line, in the succession of paragraph 4; these lines help to memorize the shapes of the constellations.

6L. Does the line through the 'pointers' α and β UMa pass exactly through the pole star? Estimate the deviation.

How far from the true pole is the pole star?

Estimate in degrees the distance between α and β UMa.

Looking at the sky, how will you be able to find the meridian where $\alpha = 0$?

When this 0^h meridian is directed towards the South and coincides with the observer's meridian, the sidereal time is 0^h. Realize how the 0^h meridian rotates in the course of one day. Estimate the sidereal time at this moment.

7L. Look up in the atlas the 7 constellations which we have studied. They correspond roughly to the following right-ascensions:

0^h	Cas	
3^h		Per
6^h		Aur
12^h	UMa	
15^h	UMi	
15^h		Boo
20^h		Cyg

To bear these numbers in mind will help in visualizing the assembly of the constellations in the sky.

8S. Finally let us return to the open air. Repeat the 7 constellations. Compare your drawing with the sky. Sketch the position of the horizon, put on your sheet the date and the hour.

Look to ζ UMa: can you distinguish Mizar and Alcor? Look through binoculars and through the telescope. Make a sketch of what you observe.

In order to save time we list the coordinates of some of the stars mentioned, as shown in Table I.

TABULATION

Name of star	Approximate coordinates		Precise coordinates		m
	α	δ	α	δ	
α UMa	$11^h\,0^m$	$62°$	$11^h\,01^m$	$61°\,56'$	$1^m.9$
—	—	—	—	—	—
—	—	—	—	—	—

Reference

SHAPLEY, H. and HOWARTH, H. E.: 1929, *A Source Book in Astronomy*, McGraw Hill, New York.
  – Quotation from Bayer, p. 21.

Preparation

For each student: a sheet of polar graph-paper, a template.

For each pair: *Norton's Star Atlas; Astronomical Ephemeris* (copies of former years can be used): a small flashlight (dimmed).

A few pairs of binoculars; a strong flashlight; telescope.

TABLE I

Name of star		α	δ	m
UMi	α	1 49	+89 02	2.0
	β	14 51	+74 20	2.2
	γ	15 21	+71 59	3.1
	δ	17 45	+86 36	4.4
	ε	16 50	+82 06	4.4
	ζ	15 46	+77 55	4.3
	η	16^h 18^m	+75° 50′	5.0
Cas	α	0 38	+36 18	2.5
	β	0 07	+58 55	2.4
	γ	0 54	+60 29	2.2
	δ	1 23	+60 01	2.8
	ε	1 51	+63 27	3.4

A2. SOME AUTUMN CONSTELLATIONS

The Problem

To become familiar with some constellations, visible in the autumn evenings. The use of star maps.

Procedure

1S. From the terrace we look at the Pole Star and estimate the course of the celestial equator. We shall now study some new constellations, connecting them to those with which we are already familiar.

We suggest studying two sectors especially:

between $\alpha = 18^h$ and $\alpha = 21^h$ Cygnus, Lyra, Aquila
between $\alpha = 21^h$ and $\alpha = 3^h$ Pegasus, Andromeda, Perseus

Each sector should be grasped as a whole, just as we study Africa, Asia, America individually. Then the connection to the surroundings should be made. Compare with our small star map, faintly illuminated.

2L. In the laboratory we prepare rectangular coordinate paper:

α from 18^h over $24^h = 0^h$ up to 4^h
δ from $0°$ up to $60°$.

Put 1.5 cm $= 1^h$ r.a., and 1 cm $= 10°$ decl. – Are these two scales in the right proportion?

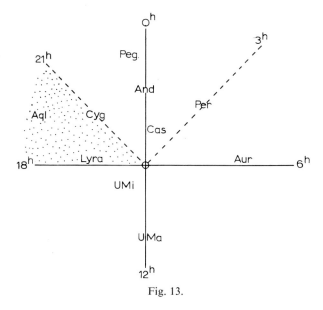

Fig. 13.

Numerate the scales. *Take care to numerate the α-numbers in the correct sense!*

3L. Plot the following stars on your coordinate paper; give them the right magnitude; draw the connecting lines.

Cygnus	$\alpha - \gamma - \beta$; $\delta - \gamma - \varepsilon$	α Cyg = Deneb
Lyra	$\alpha - \beta - \gamma$	α Lyr = Vega
Aquila	$\beta - \alpha - \gamma$	α Aql = Altair
Pegasus	$(\alpha$ And$) - \beta - \alpha - \gamma - (\alpha$ And$)$	
Andromeda	$\alpha - \beta - \gamma - (\alpha$ Per$)$	
Perseus	$\gamma - \alpha - \delta$	

To save time, we give the coordinates of the first mentioned stars (see Table I).

TABLE I

	α	δ	m		α	δ	m
Cygnus α = Deneb	20^h 40^m	+45°.1	1.3	Lyra α = Vega	18 36	38.7	0.1
γ	20 21	40 .1	2.2	β	18 49	33.3	3–4
β	19 29	27 .9	1.6	γ	18 58	32.6	3.3
δ	19 44	45 .0	3.0	Aquila β	19 54	6.3	3.9
ε	20 45	33 .8	2.4	α = Altair	19 49	8.8	0.9
				γ	19 45	10.5	2.8

Look up yourself the coordinates for Pegasus, Andromeda, Perseus, first in the *Star Atlas*, then in the *Astronomical Ephemeris*.

4L. In the constellation Andromeda the celebrated spiral galaxy is found at $\alpha = 0^h 38^m$, $\delta = +40° 50'$. Indicate this point on your map. Mark also the approximate position of the very important *vernal point*, also called: "the First of Aries". Look up in your atlas the constellation of the Fishes to which it belongs now. The hour circle through this point is taken as the origin of the R.A. coordinates; it is the same hour circle which we have already noticed in the former exercise running through β Cas.

5S. Compare the sky with your drawing and with the *Atlas*. Notice how the atlas-map should be turned when you are looking North and when you are looking South. Identify the constellations which we have selected. Find the Andromeda system, look at it also through binoculars.

6L. Take *Norton's Atlas*, and note how the surface of the sphere has been sub-divided and represented by flat drawings, without too much distortion. Note the titles of some other star-atlases, ancient and modern, which are circulated among the students or exhibited.

(7L). A professional astronomer often uses star maps. There are also cases where he has to plot coordinates himself, e.g. when a comet passes along, and ephemerides of the orbit are announced. The following Table II is an example, which you may care to plot in your drawing.

TABLE II

Comet Wilson-Hubbard 1961 d

1961	α	δ
Sep. 6	$1^h\ 59^m$	$+58°\ 54'$
28	$23^h\ 36^m$	$47°\ 58'$
Oct. 13	$22^h\ 58^m$	$38°\ 30'$
28	$22^h\ 44^m$	$31°\ 13'$
Nov. 7	$22^h\ 42^m$	$27°\ 40'$
Dec. 17	$22^h\ 59^m$	$20°\ 39'$
27	$23^h\ 14^m$	$19°\ 49'$

Circ. Bur. cent. int. Telegr. astr., nr. 1772 and 1775.

References

BEČVÁŘ: 1951, *Katalog*, Prague. BEČVÁŘ, 1956, *Atlas Coeli.*
DE CALLATAY, V.: 1955, *Atlas du Ciel*, Brussels.
SCHLESINGER, F. and JENKINS, L. F.: 1964, *Catalogue of Bright Stars*, New Haven.
SCHURIG, R. and GÖTZ, P.: *Himmelsatlas*, Mannheim, many editions.
VON BRONSART, H.: 1963, *Kleine Lebensbeschreibung der Sternbilder*, Stuttgart.

Preparation

For each student: A sheet of rectangular coordinate paper, a template.

For each pair: *Norton's Star Atlas; Astronomical Ephemeris* (copies of former years); a small flashlight (dimmed) or the lamp, Figure 9.

A few pairs of binoculars; a few planispheres; a strong flashlight; a small display of star atlases, old and new.

A3. SPHERICAL TRIANGLES

The Purpose

To formulate problems in terms of spherical triangles. To become familiar with some of the simplest formulae.

For each little problem a rough sketch of the celestial sphere has to be made. Computations are carried out by each partner, afterwards they are compared. Logarithms are used rarely: modern computing machines can handle the numbers themselves so easily. We make use of our slide rule, and neglect fractions of a degree.

Procedure (L)

1. Compute the angular distance between Deneb and Vega. Their coordinates have already been used in A2. – Repeat the computation by logarithms, adapting the formula for this purpose.

2. In order to know whether the light of a star may be weakened by interstellar clouds floating near the galactic plane, we must ascertain the galactic latitude of this star. Find this for the star Antares in Scorpius. The pole of the Milky Way corresponds with $\alpha = 12^h\ 49^m$, $\delta = 27°\ 24'$.

3. In what direction is the earth moving at this moment, due to its revolution around the sun?

To find this, look up the solar coordinates λ_0, β_0 in the A.E. Draw the plane of the ecliptic, find λ and β, then α and δ of the required point on the sphere. Look up in a *Star Atlas* where this is located. Is it actually on the ecliptic as there traced? $\varepsilon = 23°\ 27'$.

4. Compute azimuth and altitude of Sirius, observed at your observatory at $9^h\ 0^m\ 0^s$ sidereal time.

5. At what slope does the rising sun ascend with respect to the horizon? Show that this angle is equal to the parallactic angle and compute its value for your observatory. How does it change in the course of the year?

6. Compute one of these problems with a precision of a minute of arc.

7. Solve some of the problems with the astrolabe.

Example of tabulation for problem 1.

$$
\begin{aligned}
\text{Deneb} \qquad \alpha &= \ \dots\dots\dots \\
\delta &= \ \dots\dots\dots \\
90° - \delta = b &= \ \dots\dots \\
\cos b &= \ \dots\dots\dots \\
\sin b &= \ \dots\dots\dots
\end{aligned}
$$

$$\text{Vega} \qquad \alpha' = \quad \dotsb$$

$$\delta' = \quad \dotsb$$

$$90° - \delta' = \quad c = \dotsb$$

$$\cos c = \quad \dotsb$$

$$\sin c = \quad \dotsb$$

$$\alpha - \alpha' = \quad A = \dotsb$$

$$\cos b \cos c = \quad \dotsb$$

$$\sin b \sin c \cos A = \quad \dotsb$$

$$\cos a = \quad \dotsb$$

$$a = \quad \dotsb$$

References

BECKER, FR.: 1934, *Grundriss der sphärischen und praktischen Astronomie*, Berlin.
BECKER, L.: 1930, *Monthly Notices Roy. Astron. Soc.* **91,** 226. The graphs have been reproduced on an enlarged scale, accuracy 1'.
KOHLSCHÜTTER, E.: *Messkarte zur Auflösung sphärischer Dreiecke*, Reimer, Berlin.
SMART, W. M.: 1962, *Textbook on Spherical Astronomy*, chapter I.

Preparation

For each student: slide rule, trigonometric tables;
    For each pair: *Star Atlas*; *Astronomical Ephemeris*.
    A few astrolabes.

A4. SUN-DIALS

A sun-dial has a *style* or gnomon, which in general is parallel to the earth's axis; its shadow falls on a plane between *hour lines*. The simplest cases are:
  the horizontal sun-dial, designed on a horizontal table;
  the vertical sun-dial, designed on a vertical wall, facing South.

The Problem

To draw the hour lines for each of these cases.

Procedure (L)

1. Study first *the horizontal sun-dial*, either by means of a globe, or by drawing a celestial sphere.

(a) The axis of the globe is put under an inclination corresponding with the latitude of your observatory. This axis represents the *style* of the sun-dial.

(b) Take one of the hour-circles and let this be the hour-circle in which the sun is located. It revolves around the celestial axis at a uniform angular speed. (We consider here the *mean sun*.) The plane of this hour-circle contains the shadow of the style.

(c) The hour-line, therefore, is directed towards the intersection of this hour-circle with the horizon.

2. Let the globe revolve and put the hour-circle which you have selected at hour-angles t, corresponding with the successive hours. Read roughly the azimuth A of the shadow line for each hour t.

3. These azimuth angles are obtained more easily and accurately by drawing a

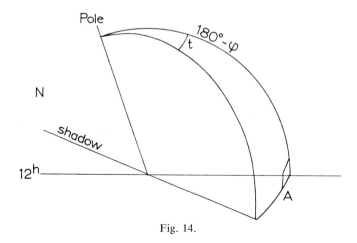

Fig. 14.

spherical triangle. The latitude φ determines the altitude of the pole, we know the mean solar time t (let us reckon it here from noon) and we notice that the triangle has a right angle. Find a formula giving directly A in terms of φ and t.

Compute the values of b for the successive hours between $t=0$ and $t=6^h$. From these, the shadow line for the other hours is directly derived. Compare with your first rough estimates.

(4.) Repeat the calculation for a vertical sun-dial.

(5.) Between the real sun and the mean sun there is a small difference $\varDelta t$, called the equation of time, which may amount to 16^m at most. What difference $\varDelta A$ in the azimuth of the hour lines would correspond to this difference of time?

(6.) Draw the hour-lines on a piece of plywood in which a cut has been made with a saw; into this slit, insert a piece of cardboard, cut it to the right elevation, and glue it to the wood. This little model may be tested!
This is a good opportunity to determine the South-North direction by a shadow line.

In exercise A6, section 5, you will find how to determine the moment of noon at your locality. Have the sun-dial prepared and fix its position.

Look at the small exhibition of ancient and modern sun-dials, books on sun-dials and photographs.

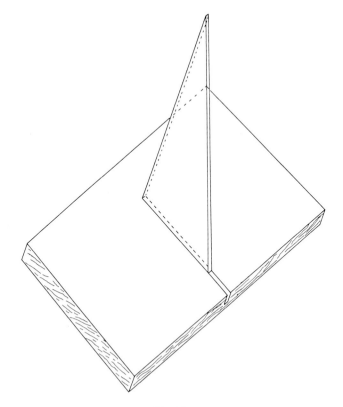

Fig. 15.

Example of tabulation

$$t=0 \quad 1^h \quad 2^h \ldots\ldots$$

$$0° \quad 30° \quad 60° \ldots\ldots$$

A (estimated)$=$ ....

$\log \tan t =$ ....

$\sin(180° - \varphi) =$ ....

$\log \tan A =$ ....

A (calculated)$=$ ....

Reference

MAYALL, R. N. and MAYALL, M. L.: 1938, *Sun Dials*, Boston.

Preparation

For each student: trigonometrical tables, 3 decimals, natural and logarithmic values.
Optional:
For each pair: a piece of plywood with a saw's cut; a piece of cardboard; scissors.
A few globes.
Display of sun-dials, books, photographs.

A5. THE DAILY ROTATION OF THE EARTH

All stars describe in one day a circle around the celestial pole. Such circles have a length of $2\pi \cos\delta$. Consequently the path which the star describes corresponds with $360° \cos\delta/24 = 15° \cos\delta$ per hour, $15' \cos\delta$ per minute.

In several cases the astronomer avails himself of this very regular motion in order to measure angular distances. It is true that the stars move along parallels, while we are in general interested in distances along a great circle. But the difference is negligible, as long as the angular distance remains small.

The Problem

To determine, for stars of different declinations, how long it takes them to cross the field of our telescope. From this, to derive the angular field of this telescope.

Procedure

1S. The daily motion of the stars is so rapid that it may be perceived almost directly, especially for stars near the equator. Looking Southward, take a stand where you see a rather bright star which is *very* near to a wall or a pole. Remain motionless. You will see very soon that the star has moved either towards the object or away from it.

2L. Note the number of your small telescope, look to see how it moves, how it can be clamped, how it is focussed. In clamping, never use your strength! Clamp gently; or *very* gently if you wish to make slight corrections later.

3S. Carry your telescope to the terrace and put its legs in the right positions, the axis pointing North; direct the instrument towards a bright star. The tube should always be on the opposite side of the column from the object you wish to observe.

(a) Look *along* the tube, above it and to the side of it; the star should be somewhere in the field now. Clamp the telescope very gently, so that its position may still be corrected.

(b) Focus. Your partner illuminates the objective from the side, so that the cross-wires become visible.

4S. Bring the star onto the intersection of the cross-wires, by making slight corrections. As soon as this has succeeded, call: 'Now!' and *do not touch the telescope any more*. Your partner reads his watch at that moment (in seconds of time).

The star moves across the field, because of the daily rotation. Look now and then through the eyepiece and watch for the moment when it reaches the boundary of the field. Then call 'stop' and let your partner read the time again.

5S. The interval t_2-t_1 is inversely proportional to the speed of rotation.

Identify the star with your map and make a note. This is very important, otherwise your observation is useless! Note also the approximate declination.

6S. Repeat this for stars of different declinations and each time determine the transit time t.

Try to find at least one star below $\delta = 30°$, another between $50°$ and $60°$, and one above $60°$.

7L. Plot, in the laboratory, $1/t$ against $\cos \delta$. A straight line through these points should pass also through the origin (Why?).

8L. How large is the angular field of your telescope? Would it change were you to diaphragm the objective? What construction element in your telescope is it, which limits the field?

You can now roughly estimate the distance between two stars, if they are together in the field.

9L. Take one of the drawings, made in the exercises A1 or A2, and insert a circle with the diameter of your telescope field. Draw, also on the same scale, a circle corresponding with the sun or the moon.

10L. We have used our watch, giving ordinary civil time. Is this allowed? If not, what time should we actually have used? Estimate the error which we have made.

Reference

DANJON, A.: 1952–53, *Astronomie générale*, Paris, see pp. 42–43.

Preparation

For each pair: Experimental telescope; *Star Atlas*; watch, marking seconds.
For each student: Rectangular coordinate paper.

A6. CONVERSION OF TIME

We consider only:

Sidereal time ST, reckoned from 0^h at upper transit of the First of Aries (sidereal noon). GST and LST correspond to Greenwich ST and local ST.

Universal time UT, reckoned from 0^h at lower transit of the mean sun (solar midnight) for the Greenwich meridian.

Standard time, as used in civil life, differs from UT by a whole number of hours, according to the country.

For simplification we shall neglect the small difference between ephemeris time and universal time.

The conversion exercises must be carefully prepared, since the arrangement of the *Astronomical Ephemeris* has been repeatedly modified in recent years. The tables are so skilfully arranged, that time conversion becomes very easy.

Conversion of time into arc. (*Astron. Ephem.* Conversion Tables XI and XII.)

1L. Note once and for all the longitude of your observatory expressed in arc (W. Longitude is positive); you will need this often. Convert this longitude into time (full precision).

Reconvert the longitude, expressed in time, to angular measure and see whether you find exactly the number from which you have started.

PROBLEMS AT GREENWICH

2L. Convert an interval of ST into UT. (Conversion Tables VIII and IX.)

3L. Convert UT data into ST. (Table 'Universal and Sidereal Times'; Conversion Tables VIII, IX.)

On the day selected:

$$0^h \text{ UT corresponds with } \ldots\ldots\ldots \text{GST}$$

$$t^h \text{ (UT) is equal to} \begin{cases} t \text{ (ST)} \\ + \varDelta t \text{ (ST)} \end{cases}$$

$$t^h \text{ UT corresponds with } \ldots\ldots\ldots \text{GST}$$

t (UT) or t (ST) designate time *intervals* in UT or ST. The *Astronomical Ephemeris* gives tables for $\varDelta t$.

An alternative method may be followed, using the column 'Transit of First Point of Aries'.

PROBLEMS INVOLVING THE GEOGRAPHIC LONGITUDE

The systematic solution is found via the time data for Greenwich, which are listed in the Ephemeris.

We distinguish Greenwich sidereal time GST and local sidereal time LST.

4L. For an observatory at W. longitude L, convert $20^h 0^m 0^s$ UT into local sidereal time LST.

$$UT \rightarrow GST \rightarrow LST$$

To transform the standard time of the country where the observatory is located into UT: add a whole number of hours wl (roughly equal to the WL of the country, expressed in hours).

5S. Determine the direction NS by observing the shadow of a plumb line at real noon. This quick method is often used when instruments have to be roughly orientated, e.g. on an expedition. A precision of half a degree is easily reached.

In the *Astronomical Ephemeris* look up the tables on the sun. Column *Ephemeris Transit* gives the moment a of transit at Greenwich in ephemeris time, which for most purposes is equivalent to universal time. Noon at your observatory will occur at a moment $a + $WL in universal time. Interpolate a for that moment, between two successive Greenwich transits.

$a + $WL in universal time corresponds to $a + $WL $-$ wl on a standard time clock. The difference $\boxed{\text{WL} - \text{wl}}$ is known once for all (compute!). Investigate whether the interpolation of a is really necessary within the limits of the precision which we have required. If not, to what does the operation reduce?

Now do it yourself! Determine the NS direction on the campus or in your garden by means of a plumb line.

Preparation

For each pair: *Astronomical Ephemeris*, by preference 1965, 1966, or following years.

Demonstrations

On whole hours, many radio stations send a signal of 'six pips', separated by intervals of 1 sec; the last pip coincides with the exact hour. The correction of the UT clock is determined with a precision of $0^s.1$.

Find a suitable place where you can hear both the beats of the UT clock and those of the LST clock. Listen to the moments of best coincidence. Notice the shorter duration of the seconds of ST and the gradually increasing phase difference.

A7. SIMPLE MEASUREMENTS WITH AN ALTIMETER

Instrument

To measure the altitude of a star when on an expedition, one uses a theodolite, a portable transit instrument, or a sextant (on a ship). We shall avail ourselves of the very simplest means: a rectangular board, fastened to a heavy stand, which we place in the proper orientation. The board turns around a screw and its long side is pointed to the star by means of the sights M and N, then the board is gently clamped in place by a winged nut.

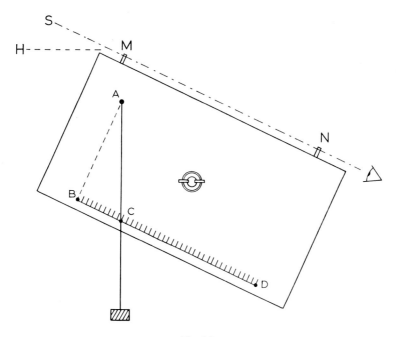

Fig. 16.

A plumb line is our reference vertical. Angle $SMH = BAC$, the altitude of the star, is measured by its tangent BC/AB; for simplicity $AB = 20.0$ cm. By comparison with a theodolite the millimeter scale BCD has been so adjusted, that the plumb line passes through the zero point B when the sightline is directed to the horizon.

Compare to altitude measurements with our refractor (A11); the precision of the pointing is much less because there is no telescope; however the angular position readings are better. Each measurement should be repeated several times. Remember Tycho reaching a precision of $1'$. And beware of the wind!

Preliminary

Make a few rough estimates of altitude without any instruments:

(a) extend your arm, spread your fingers: the distance from the end of your thumb to the end of your little finger corresponds to about 20°;

(b) stretch your arm and look at your thumb; its thickness corresponds to about 2°;

(c) try to look at the zenith and notice the stars there; then turn around over 180° and try again. You will be surprised in noticing the difference between your two estimates!

Problem I

To measure the altitude of the pole by means of the Pole Star. This is equivalent to a measurement of our latitude.

PROCEDURE

1S. Measure the altitude of the Pole Star: this is already a rough approximation for the elevation of the Pole itself. For a better determination, record the time by your watch (precision: 5 min.).

2L. Compare your watch with the clock and find, at what LST your observation was approximately made.

3L. We note, that α UMi is distant from the Pole by 70′, and that it describes a circle with such a radius in 24^h ST. Let t be the hour angle of the Pole Star, then $\varphi = h - 70' \cos t = h + \Delta h$.

Thus by adding to the observed h the 'correction' Δh, we find the true polar altitude φ.

In the *Astronomical Ephemeris* Δh will be found tabulated as a function of LST.

Apply the correction and find a better value for φ. Compare with the results of other students.

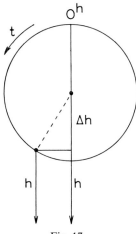

Fig. 17.

4L. The observation of the Pole Star is also a simple method for finding the direction of true North. How should our formula be modified in that case?

Preparation

For each pair: altimeter; watch; *Astronomical Ephemeris* (copies of former years can be used); LST clock.

Problem II

On an expedition, if no radio signals are available, it is often useful to have our own time determinations. By measuring the height of a star which is rising or setting we can easily find its hour angle and the sidereal time. It can be shown that this measurement is the most precise when the star is on the prime vertical (W or E).

5L. Adjust your watch to the astronomical clock which gives LST (precision: $0^m.1$). The clock correction has been ascertained and should be applied.

6S. Select a suitable star, more or less in the E or W direction and (if possible) not high above the horizon. Determine several times successively its height with the altimeter. At those moments when the pointing is correct your partner notes the time on his watch (precision $0^m.1$). It is striking how quickly the height changes; the difference is already noticeable after 1^m or 2^m.

7L. Plot the readings of the altimeter h against the time t, draw a smooth mean curve and select one of its points for further calculation.

8L. Derive the LST from the spherical triangle Pole-Zenith-Star. Compare with the observed time as it was read from your watch, taking into consideration that this loses 10^s in one hour with respect to LST.

References

CHAUVENET, W.: *A Manual of Spherical and Practical Astronomy*. Philadelphia, Vol. I, ch. 5.
NIJLAND, A. A.: 1903, *Astron. Nachr.* **160**, 257.

A8. ECCENTRICITY OF GRADUATED CIRCLES (L)

We investigate all kinds of graduated circles, mostly parts of portable instruments. Old instruments are especially suitable, because there the errors are more apparent.

A graduated circle is a highly valuable component of astronomical instruments, which has to be handled with care. Such a circle is usually silvered. To avoid oxidation, it should not be touched with the fingers.

1. Examine first how the circle on your instrument is constructed. In general there are two verniers, often connected by an alidade. Is it the circle which turns or the alidade?

Clamp the alidade and use the correcting screw. Are the microscopes or magnifiers well focussed?

The scale should be well illuminated; avoid lateral asymmetry: it makes the graduation appear more distinct, but will slightly shift the apparent position of the graduation; use a desk lamp or a flashlight.

2. Look now at the graduation. Find the value of a division and see what is the meaning of the numbers. Make a rough sketch of the graduation over a few degrees. The vernier will be studied later on.

The Problem

The centre of the alidade never entirely coincides with the centre of the graduated circle. This has the effect that the direction of the alidade, which should be read z, will be erroneously read z'. We have: $z - z' = e \cdot \sin(z - E)$, where e = eccentricity = distance between centre of circle and centre of alidade. E is the position where z' becomes equal to z. The error fluctuates between $+e$ and $-e$; ordinarily it is very small. Our problem is to determine e and E.

Procedure

3. Put one of the verniers very precisely with its zero division on the zero-point of the graduation; if the adjustment is correct, the position of the neighbouring vernier divisions to the right and to the left will be symmetric. Your partner now reads the position of the opposite vernier, looking sharply and without prejudice. For the moment he only determines at how many divisions n of the vernier to the right or to the left complete coincidence is observed; sometimes he will read in halves of such a unit. (The value of these units will be found later.)

4. Repeat this observation at the positions $30°, 60°, 90° \ldots 360°, 30°, 60°$. Do not try to put a systematic trend in your numbers, simply continue. By the repetition of a few readings you will have some estimate of the precision attained. It would also

be worthwhile to make a few readings with intervals of only 1°; this gives an impression of the irregular errors of the graduation.

5. Plot n as a function of z. You get a better view by repeating the points 30°, 60° ... beyond 360° at 390°, 420°, ...

In general, the points spread irregularly because of reading errors and irregularities in the graduation, but usually a tendency towards a sine line will become apparent.

6. If the two zero readings of the verniers are not separated by precisely 180° but by 180° + α, the sine curve as a whole will be located higher or lower than the horizontal axis. Actually $\bar{n} = \alpha$. Compute α (in the same units as n) and draw the corre-

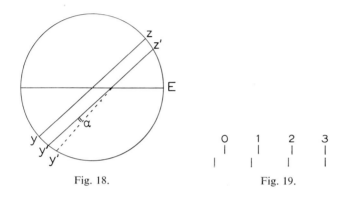

Fig. 18.   Fig. 19.

sponding horizontal axis of the sine line. Now draw the sine line itself between the observed points, giving attention to the distance between the maximum and the minimum, between the intersections with the axis, etc.

The amplitude of the sine curve corresponds to the very small angle e/r, where r is the radius of the alidade. From the zero points, read E.

7. We must now study our verniers more closely. Put the zero line on one of the divisions of the circle. From there on there is an increasing divergency between the vernier and the circle. Look to see after how many divisions coincidence is again observed. This gives the difference between a unit of the one and a unit of the other division. You will understand now the numbers on the vernier scale. Draw a part of both scales (Figure 19).

If you put the vernier in an arbitrary position, you can first roughly estimate the reading, then obtain a more precise value from the vernier.

8. The value α and that of e/r can now be expressed in angular measure, then in radians; αr and e are found in fractions of a millimeter.

(9.) The sine curve is determined more accurately by least squares. One finds:

$$n \cdot \cos z = -e \cdot \sin E$$
$$n \cdot \sin z = \quad e \cdot \cos E.$$

These equations are easily solved.

Reference

CHAUVENET, W.: *A Manual of Spherical and Practical Astronomy*. Philadelphia, Vol. II., ch. 2.

Preparation

For each pair: a graduated circle, by preference from an old instrument; 2 verniers with magnifiers; desk lamp or flashlight; a sheet of rectangular coordinate paper.

A9. THE ADJUSTMENT OF A TELESCOPE

The Problem

We want to check the orientation of our experimental telescope, which has an equatorial mounting. When putting a telescope in operation, it is necessary to mount it in the right position, so that the coordinates of the stars may be directly read on the graduated circles. The essential requirement is that the polar axis should be directed exactly at the pole.

Procedure

1L. First look at your small telescope in the laboratory. We shall refer, in future, to the vertical *column* and the (*polar*) *axis*. We imagine that it has been placed on its stand, the polar axis pointing towards the North Pole. Notice how the tube moves in declination and in hour angle. How are the circles graduated? What is the value of each division? Point the telescope to $\delta = 90°$, to $\delta = 30°$, to $\delta = 0°$. Then point it successively to several hour angles and notice the graduation of the hour circle. The graduation has no + or − sign; use your common sense!

Always estimate the angles first very roughly, then read the graduation, in order to avoid confusion of sign.

2S. The telescope is now put into position on the terrace: one leg fits into a hole, the second one in a groove, the third one rests on a little plate. There are no degrees of freedom left.

We shall first have to check whether the declination circle actually reads 90°, when the telescope tube is parallel to the axis. – Point the telescope to a bright star M_1 near the meridian and read its position m on the declination circle. Keep the telescope clamped in declination and turn it over 180° in hour angle; if it was first to the right

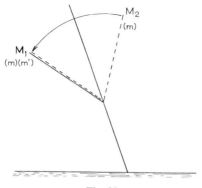

Fig. 20.

of the column, it will now be to the left. It is now directed to a point M_1 of the sky, symmetric to the star with respect to the polar axis; the reading has not changed. From this position, point again to the star by changing the declination and read again the graduation (m'). The mean of the two declination readings m, m' gives the precise direction of the axis. Check whether this coincides with the division 90° of the graduation; or determine the amount $\varDelta$ of the correction, to be applied to all declination readings. If necessary, press a little pointer of tape on the circle, at division $90° + \varDelta$, indicating the position where the telescope is parallel to the axis.

3S. Now let us check how the telescope has to be orientated. A preliminary position may be found, either with the shadow of the plumb line at noon, or with the Pole Star at night. In our case rather precise adjustments of the telescope stand have already been made.

A very simple test can be made even without graduated circles: a star, taking part in the daily rotation of the sky, moves along a parallel; consequently it should remain at the centre of the field while our telescope, clamped in declination, rotates in hour angle around the polar axis.

Of course we select stars which are not too near the pole. A star at the south side of the sky will give a check on the azimuth of the polar axis. A star at the east or west side will give a check on the elevation of the axis.

Shift the telescope slightly out of position by inserting a metal strip under one of the legs (the leg resting on a flat disk). Notice the deviation and assess how far the method is sensitive.

4S. A quicker method is available since our instrument has a declination circle. It has the advantage of being independent of time measurements. – Observe a number of well-identified stars at different hour angles. Read each time the precise declination and note very roughly the hour angle.

5L. Compare the observed declinations with the declinations in the *Astronomical Ephemeris*; plot the differences against the hour angle. In general you find a sine line. This simple graph gives at once complete information about the position of the telescope axis. The height of the mean horizontal line corresponds to the zero point error of your declination circle. The hour angles where the difference $\delta_{obs} - \delta_{true}$ reaches its extreme value inform you of the direction towards which the axis deviates from the true pole. The amplitude measures the amount of the deviation.

6S. If the orientation of the telescope has proved sufficiently precise, insert a metal strip under one of the legs and repeat the measurements. Again assess the sensitivity of the method.

TABULATION

Name of star	hour angle	δ (obs.)	δ (Ephem.)	$\varDelta\delta = \text{obs.} - \text{true}$
....	....	....	....	....
....	....	....	....	....
....	....	....	....	....

(7). Actually we would have to check also the zero point of the graduation in hour angle. This will be done in the next exercise (A 10, § 6).

Reference

Sidgwick, J. B.: 1954, *Amateur Astronomer's Handbook*, London, chapter 16.

Preparation

For each student: rectangular coordinate paper.

For each pair: the experimental telescope; a metal strip, 3 mm thick; *Astronomical Ephemeris*; star map; flashlight.

A10. TO POINT THE TELESCOPE TO A STAR

This simple operation is at the basis of almost any astronomical observation and should be repeatedly practised.

We shall operate in two steps:

(1) first we test whether the support of the telescope has the right orientation and whether the scales are correct;

(2) then we put the telescope in a predetermined position and check whether the star is in the field.

Procedure

PROBLEM I

1L. For these observations we make use of the astronomical clock, giving local sidereal time LST. Apply the clock correction and regulate your watch as near as possible; if there remains a difference, it should in any case be smaller than 30^s and your watch should be in advance compared to the sidereal clock. We now shall assume that it shows local sidereal time during the whole evening.

2L. As in the former exercise, notice how the telescope may be clamped and how the graduations run. Direct the telescope to an imaginary star and read the coordinates.

3S. Select a bright star and direct the telescope to it. Clamp your instrument very gently. Your partner illuminates the objective from aside, so that the cross-wires become clearly visible. Now slightly correct the telescope position till the star is precisely in the centre of the field. Give the signal 'now!'.

4S. Your partner reads his watch. Then, more leisurely, you read the hour angles and the declination on the graduated circles. Always first estimate the coordinates very roughly by eye, in order to avoid errors of sign!

Carefully identify the star by means of the *Star Atlas*.

5S. Repeat this for another star, in another part of the sky.

6L. Look up the coordinates of your two stars. Compare the hour angle and declination, determined by you, with the values from the Ephemeris (h.a. = local sidereal time $-\alpha_*$).

If the agreement is reasonable, proceed to the second part of the exercise. If not, warn the instructor, who might be able to correct at once the telescope errors by slightly rotating the hour circle.

PROBLEM II

7L. We take one of the same two stars which you have already observed and for which you know α and δ from the Ephemeris. However, this time we shall have to put the telescope in a predetermined position and then check whether the star is in the field.

Estimate how much time you will need to adjust the telescope, say 20 min. Thus, you will have to be ready for observation 20^m hence. We want to put the telescope in such a position that the star will *then* be in the field.

8S. Adjust your telescope to this position. Then wait till the star crosses the field and reaches the centre. Note the sidereal time. If necessary, correct the declination quickly and read the improved value.

9S. Repeat the operation with other stars.

From a comparison of the results, some conclusions about the accuracy of your telescope may be drawn. Note therefore the number of the instrument which you have used.

Note. In the course of the evening your watch will have deviated from the sidereal time by about half a minute. This is practically negligible, compared to other sources of error.

Preparation

For each pair: experimental telescope; flashlight; *Star Atlas*; *Astronomical Ephemeris*.
   Astronomical clock, giving LST.

A11. FUNDAMENTAL MEASUREMENTS WITH
THE MERIDIAN CIRCLE

Introduction

The meridian circle is a fundamental instrument of position astronomy. We shall see how some important data are obtained with this instrument. We clamp our experimental refractor at an hour angle of precisely 0^h, and shall now use it as a 'meridian circle'. Of course there is no comparison between the stability and the precision of a real meridian instrument (which moreover can be reversed) and our primitive model.

For fundamental measurements we would have to determine simultaneously: (a) the position of the pole and the declination of stars; (b) the local sidereal time and the right ascensions. This would require repeated measurements, spread over a considerable time, and is seldom carried out by the astronomer. In ordinary practice he assumes that some data are known and he determines the other ones.

Problem I

(1S.) *To determine the horizontal.*

Measurements with the meridian circle always begin by a determination of the horizontal position of the telescope; in practice this is done by a mercury collimator. In our case we use the level which is fixed on the tube of our instrument. Probably it is sufficiently parallel to the optical axis; but, if desired, an exact test may be obtained in the following way.

General plan: (a) put the telescope parallel to the polar axis; (b) then put the level parallel to this axis; suppose there is a difference Δ between the two positions. This being determined, we bring the level in the horizontal position and correct by an amount Δ: the telescope is now horizontal.

(a) The first operation has already been done in exercise A9. Put the telescope on its stand, clamp it at an hour-angle 0^h and direct it towards any distant mark M_1; read the declination circle (m) and clamp it. Turn over $180°$ in hour-angle; the telescope is now directed towards a symmetrical point M_2. Direct again towards the same mark from which we started; the reading is now m'. The mean of the readings m and m' gives the position where the telescope is parallel to the polar axis (see Figure 20, p. 27.)

(b) At hour-angle 0^h bring the level in the horizontal position H_1 (our 'mark' is now the horizon), and read the declination circle (n, Figure 21). Turn over $180°$ in hour-angle, the level is now directed along H_2. It would seem logical to bring it back towards the horizon H_1. However, a spirit level has to be, of course, *above* the telescope; we therefore modify the method and move the telescope until again the horizontal position H_1' of the level is reached. The reading is n'. It is clear that the mean of the readings n, n' at H_2 and H_1', minus 90°, is the position where the level is parallel to the polar axis. Is there a difference between the two results under (a) and (b)? If so, let us call it Δ.

When therefore the level is horizontal, the telescope will still be slightly inclined; we shall have to turn over a small angle Δ in order to have it pointed at the horizon. (Take care to give Δ its correct sign!)

Note once for all the horizontal position of the telescope.

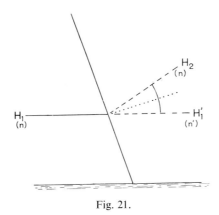

Fig. 21.

Problem II

Assuming your latitude and the LST to be known, to determine the coordinates α, δ of a star.

PROCEDURE

2L. Adjust your watch to the LST clock, taking into account the clock correction. Note the remaining watch correction (in seconds). Look in the *Ephemeris* and select a rather bright star which will culminate half an hour hence. Note the approximate declination and identify the star on your Atlas. Compute the elevation at which the star will transit.

It should be clear that we have consulted the *Ephemeris* only for a preliminary orientation, but that we have to find independently the stellar coordinates from our own observations.

3S. Put your 'meridian circle' in its fixed stand on the terrace. Now clamp the telescope gently at the approximate elevation where the star is expected to cross the field. Your partner manipulates the lamp and is ready to give a faint illumination to the field, in which the cross-wires then become visible.

4S. As soon as the star appears in the field, correct the elevation of the telescope so that the star runs along the horizontal cross-wire, and clamp it more strongly.

At the moment when it crosses the vertical wire, give the signal 'stop'. Your partner first reads the time (in seconds) and then notes carefully the elevation.

Check again whether you have identified the star correctly.

5L. Compare again your watch with the LST clock, and again note the watch correction. Interpolate roughly this correction for the moment of your observation.

6L. Now from your notes derive α and δ. Compare with the *Ephemeris* coordinates.

Problem III

Assuming the coordinates α, δ of the stars to be known from the *Ephemeris*, to determine LST and the altitude of the pole (= latitude).

PROCEDURE

7L. Our previous determination can be used in the reverse sense. We find the correction of the LST clock and the elevation of the pole above the horizontal. If all operations have been carefully carried out, the pole is found at reading 90° of our declination circle, since this has already been checked.

These are methods commonly used. However, with a precise instrument, several corrections would have to be applied.

Problem IV

To check the rate of your watch.

PROCEDURE

Our experimental telescope, *carefully clamped* in the position used for the star transit, is put aside till the following day, when it is returned to precisely the same position on its stand as before.

8S. Prepare for observation of the transit at the same LST. Repeat the observation of the previous day.

Take into account the difference between the length of the day in ST and in UT, amounting to $3^m 57^s$. Then, from your notes, check the rate of your watch.

Reference

OLBERS, J. G.: 1801, *Monatliche Korrespondenz* **3**, 125; quoted by K. Schwarzschild in *Neue Beiträge zur Frage des Mathematischen und Physikalischen Unterrichts* (ed. Klein and Riecke), Teubner, 1904.

Preparation

For each pair: experimental telescope; *Astronomical Ephemeris*; *Star Atlas*; flashlight; watch, marking seconds.

LST clock, clock correction.

Demonstration

Passage of a star, observed by a meridian instrument.

A12. REDUCTION TO THE MERIDIAN

The Problem

Any meridian instrument deviates a little from the theoretical position however small the deviation may be. We want to find the three main constants by which this deviation is determined: the azimuth constant, the level constant, and the collimation constant.

Procedure

1L, 2S. Follow the programme, described in A sections 3 and 4, and determine the transit time of a star.

3S. Repeat this determination of the transit time for at least 4 stars at different elevations, their declinations varying over a wide range. Include also stars at the N-side of the sky.

4L. Compare again your watch with the LST clock, note again the correction. Estimate this correction by a rough interpolation for each of the transits observed, and reduce the transit times to LST.

5L. Compare these transit times to the values of α according to the *Astronomical Ephemeris* and compute Δ = observed transit time $-\alpha$.

6L. All these differences ought to be zero if the instrument and the observer were perfect. Plot Δ as a function of δ. Draw a smooth line through these points. (For stars at lower culmination, change δ into $180° - \delta$ and α into $12^h + \alpha$.)

7L. According to Bessel, this curve may be represented by the formula:

$$\Delta = m + n \tan\delta + c \sec\delta.$$

Choose three points of the curve sufficiently distant from each other; write the three equations with the three unknowns and solve. Check whether another point fits reasonably well in the curve. The constants m, n, c are expressed in seconds of time; c is the collimation constant; m and n yield the two other instrumental constants:

$$\begin{cases} a = m \sin\varphi - n \cos\varphi \\ b = m \cos\varphi + n \sin\varphi \end{cases}$$

Reduce a, b, c to angular measure. (See also tabulation.)

(Actually one would have to use the method of least squares, applied to *all* stars observed. – Consider in which part of the sky the several constants may be best determined.)

TABULATION

Before the observations:	After the observations:
LST clock	LST clock ....
LST corrected	LST corrected ....
Watch	Watch ....
LST – watch	LST – watch ....

Name of transiting star	....	....	....	....
δ	....	....	....	....
α	....	....	....	....
Observed transit time (watch)	....	....	....	....
Observed transit time (LST)	....	....	....	....
$\Delta = \alpha -$ obs. transit time	....	....	....	....

A13. THE SEXTANT

The sextant is an ingenious, simple instrument, by means of which angles may be measured with a precision of 10″, even when the observer is on a swinging ship. With it the sailor finds his position, or the time, or the distance to an object of known height.
Do not touch the silvered scale with your fingers!

The Instrument

1L. First examine the instrument. A fixed sector of 60°, light but strong, carries an *index arm* which revolves around a pivot in the centre *C* of the sector. This arm carries a small mirror; the position of the arm and of the central mirror is read on a scale by means of a vernier *V* and a magnifier. Notice how practical the arrangement of this vernier is and remember the value of one division. The light, reflected by the first mirror, reaches a second, fixed mirror *M* (*horizon glass*) and finally the

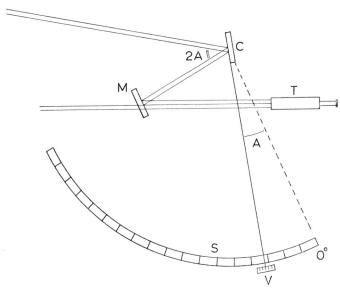

Fig. 22.

eye or a small telescope *T*. Look, also, at the coloured glasses (*shades*); notice how the index arm may be clamped and adjusted by means of the tangent screw.
When inserting the small telescope, avoid damage to the fine screw thread: always turn in the reverse direction first, a click is heard when the screw thread fits.
In making sextant observations, it is much easier to keep the image in the field if

you can lean against a wall or a pillar. Do not try to imitate the sailor who measures easily on board a rolling ship!

2S. Remove the telescope. Place the central mirror one or two degrees from the zero position. Keep the plane of the sextant vertical and look with one eye towards the horizontal skyline of a building. You see two images: (1) a direct image, observed through the non-silvered half of the horizon glass; (2) a doubly reflected image, formed by rays which have reached the eye via the two little mirrors. These two images almost coincide. Incline the sextant somewhat more, somewhat less: the coincidence persists.

3S. Rotate the central mirror over an angle A. The direct image remains, the doubly reflected image moves over an angle $2A$. You see both images superimposed, at least in the centre of the field. The top of a chimney will perhaps coincide now with the roof of the house. Two objects of which the images are thus brought to coincide, are distant by an angle $2A$; the scale graduation mentions $2A$ (and not A!).

Bring into coincidence the top of a tower with the roof line of a building below; or the top of a church-steeple with the centre of the church-clock. The directly observed object should always be the lowest of the two (why?).

Repeat the observation by means of the small sextant telescope. The coincidence is now observed over the entire field, not only in the central part. Why? – (*Never point the telescope to the sun: danger!*)

Let the instrument revolve around the sight line by a small angle to the right or to the left. The doubly reflected image describes an arc; the two objects should just touch when they pass each other.

4S. Measure also an angle in the horizontal plane: e.g. the distance between two church-steeples.

The Problem

To determine the apparent diameter of the sun and the index correction.

Procedure

5S. *Insert the necessary dark glasses in both light beams.*
First test which glasses are suitable by looking through them directly at the sun, without the telescope and holding the sextant in a skew direction.

Put the index arm approximately at the zero position again, and aim at the sun. Now let the two images touch as precisely as possible:

(a) upper limb of the reflected image on lower limb of direct image;

(b) lower limb of reflected image on upper limb of direct image. Let the readings be r and r' (in the last case you find $359°$ ..., from which you subtract $360°$). If R is the reading when the images coincide, and s the apparent diameter of the sun, we have:

$$r = R + s$$
$$r' = R - s.$$

Whence $R = \frac{1}{2}(r+r')$ and $s = \frac{1}{2}(r-r')$.

The *correction* to be applied to any subsequent reading is therefore: $-R$.

In order to obtain the index correction and the solar diameter with precision, several measures should be taken and averaged.

Compare the apparent diameter of the sun to the ephemeris value.

TABULATION

r	r'	$R = \frac{1}{2}(r + r')$	$s = \frac{1}{2}(r - r')$
–	–	–	–
–	–	–	–
–	–	–	–
		$\bar{R} =$	$\bar{s} =$

Reference

NASSAU, J. J.: 1932, *A Textbook of Practical Astronomy*, New York.

Preparation

For each pair: sextant.

A couple of copies of the current *Ephemeris*.

A14. FINDING YOUR POSITION AT SEA

(Sumner 1843 – St. Hilaire)

The Problem

After the invention of the ship's chronometer, the determination of the position of a ship became a classical operation, which is regularly applied even today, though it is now complemented by the reception of radiosignals.

For our exercise we read the UT from our astronomical clock, here replacing the ship's chronometer.

It is sufficient to measure the altitude of two stars (or a planet, or the sun) at a given time. The sailor has a free horizon and uses his sextant. For our purpose we can use the simple altimeter; illuminate the sights by means of a flashlight. Alternatively, take our telescope, placed on a more or less horizontal plane; the polar axis should be directed to the azimuth of the star, and the altitude is measured with respect to the level.

Measurements

1L. Adjust your watch to UT, by comparing with the astronomical clock; take account of the clock correction.

2S. Select a bright star and determine its altitude. Note immediately the UT time.

3S. Repeat this with another star, of which the azimuth differs by about 90° from that of the first.

4S. In order to simplify our calculations slightly and to reach a somewhat higher accuracy, these measurements are made several times during a period of about 20 minutes, alternatively on the first and on the second star. By graphical interpolation the height of the two stars is determined for the same moment of time.

(5S.) If time allows, take three stars instead of two.

Calculation (Figure 23)

6L. We assume that your *dead-reckoning position* Z is roughly known: latitude φ and W, longitude L. Choose their deviation from the true position of your place; say about 2° in each coordinate. (Such a big difference will seldom occur in practice.)

The *terrestrial projection* S of your star is determined by $\varphi^* = \delta$ and $L^* = $ hour-angle t^* with respect to Greenwich $= \text{GST} - \alpha$. Since you have found its zenith distance $z = 90°$ – the measured altitude, you know that you are on the dotted circle, the *position circle*.

Convert the moment of observation from UT into GST (exercise A6), compute L^*.

7L. The dead-reckoning position Z would have required, instead of a zenith distance $z = SJ$, a somewhat different zenith distance $z' = SZ$. (Calculate this from the

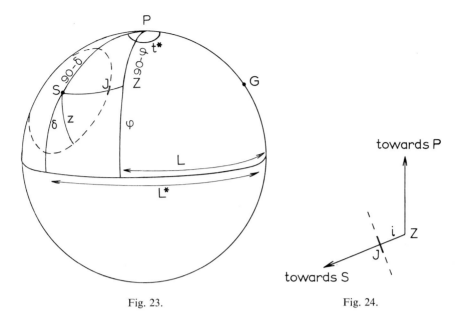

Fig. 23.   Fig. 24.

triangle *SPZ* (note that the angle *SPZ* is equal to $L^* - L$). The difference $z' - z$ is called the *intercept i*.

We wish to draw a small part of the position circle near Z. For that purpose we have to know the angle *PZS*, which actually is 180° – the azimuth of *S* as observed from Z. Compute this angle from the same triangle *SPZ*.

8L. Now reproduce the arcs *ZP*, *ZS* by straight lines on ordinary rectangular coordinate paper (1° = 2 cm). This is a sufficient approximation in the direct vicinity of Z. Measure on scale the intercept, find the intersection *J* and a portion of the position circle (Figure 24).

9L. Repeat this construction for the second star, if possible also for the third.

10L. Read the coordinates of the intersection of the position circles. Find on a map of your country to what position it would correspond and compare with the actual location of your observatory.

Reference

SMART, W. M.: 1943, *Handbook of Sea Navigation*, London, ch. 6.

TABULATION

Name of the star	α	δ	UT	GST	h	z	z'	i
1.	–	–	–	–	–	–	–	–
2.	–	–	–	–	–	–	–	–

	Star nr. 1	Star nr. 2
$\sin PS$	–	–
$\cos PS$	–	–
$\sin PZ$	–	–
$\cos PZ$	–	–
$\cos SPZ$	–	–
$\sin SPZ$	–	–
$\cos SZ$	–	–
$\sin SZ$	–	–
$SZ = z'$	–	–
$z - z' = i = JZ$	–	–
$\sin PZS$	–	–

Preparation

For each student: rectangular coordinate paper.

For each pair: altimeter or experimental telescope; flashlight; *Star Atlas*; *Astronomical Ephemeris*.

Astronomical clock, giving UT.

Note. If we wish to measure the elevation of the sun, we can apply a much more precise method. We set up an *artificial horizon*: a piece of dark glass, minimum size 10 cm × 10 cm. Adjust it by means of a water-level and small wooden wedges or pieces of paper till it is horizontal. With the sextant we measure the angle between the sun and its reflection in the glass, which is twice the altitude above the horizon. – The dark glass must be tolerably flat; glasses used for welding are often very satisfactory. There will be perhaps some difficulty at first in adjusting and directing the sextant till the two images are in the field; make a preliminary observation without the little telescope and with medium dark glasses, then insert the telescope and darker glasses. – In the case of the sun we need one measurement in the morning and one in the afternoon.

A15. PARALLAX AS A MEASURE FOR STELLAR DISTANCES

The Problem

Distances in astronomy are usually so enormous, that they can only be measured by means of large instruments and a very careful measuring technique. We shall imitate the method of working on an enormously enlarged scale and with artificial light-sources.

We select a quiet avenue allowing a free perspective over a distance of a kilometer or so. A street light at the other end represents a 'background star' A. Somewhere closer (between 200 and 400 m) we put a flashlight, covered by a green glass (to make it easily identifiable), at one side of the road. This will represent 'the star S', of which we have to determine the distance. Standing at one end of the avenue, we observe both distant light sources A and S. A few steps across the avenue, from one sidewalk to the opposite one, will show at once the parallactic shift; the star which seems to move (relatively) in the same sense as our eye is always the most distant of the two. These few meters represent the distance over which the earth has moved in the course of half a year: 300 million km.

Always take into account that the angles are very small, never use sines or tangents, but always the angles themselves or their value in radians.

Procedure

1. Put a mark W on one sidewalk. From there, with the sextant, measure the angle AWS, in minutes of arc. (Take the mean of two measurements. Do not worry about the zero correction of your instrument.)

2. Move towards the other sidewalk and put there a second mark W'. From there, determine the angle $AW'S$, as before (two measurements).

3. Measure the distance WW' between the two marks with a rope and a meter-rule.

4. The (annual) parallax p is the angle at the star, subtended by the radius of the earth's orbit. Let us first assume that our background star A is at an infinite distance away. Compute the parallax of W:

$$p = \tfrac{1}{2}WSW' = \tfrac{1}{2}(AWS - AW'S)$$

(if S is seen in both cases to the right of A, or to the left of A). Convert p into radians.

5. Let us now take into account that the background star is not infinitely distant. From statistical considerations we may have estimated that its distance is say, 900 m. Show that we shall have to apply to the angle WSW' a correction of $(WW'/900)$ rad.

Should it be added or subtracted?

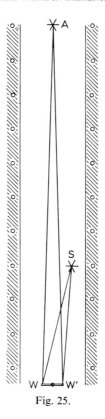

Fig. 25.

6. Compute the 'distance of the star'

$$= \frac{\text{'radius of the earth's orbit'}}{p}.$$

Compare with the direct measurement WS, carried out with a rope, 10 m in length. How big is the relative error?

Our experiment corresponds to the simplest case: the stars A and S in the plane of the ecliptic; observations made at two moments of the year where $\lambda - \lambda_0 = +90°$ and $-90°$. Walking in a circle instead of crossing the avenue would represent observations made during the whole year.

Note that the parallax of the nearest star (Proxima Centauri) is about one thousandth part of that which we measured in our experiment.

TABULATION

$AWS=$ .... $WW'=$ ....

$AW'S=$ .... $\dfrac{WW'}{900}=$ ....

$p=\frac{1}{2}(AWS-AW'S)=$.........

p (corrected)$=$ ....   Distance of $S=$ ......

Preparation

Flashlight, with a green glass; a rope, 10 m in length; a meter-rule.
  For each pair: a sextant.

A16. PRECESSION, ABERRATION, NUTATION

The Problem

In the *Astronomical Ephemeris*, you find for a number of bright stars the *mean position* and also the *apparent position*. (Since 1957 these last data are published in a separate volume.) The apparent positions vary in the course of the year. What is the reason? Or better: What are the reasons?

Procedure (L)

1. For historical reasons we select the star, studied with such great success in 1725–1728 by Bradley: γ Dra, $\alpha = 17^h 55^m$, $\delta = 51° 29'$.

Plot the mean positions at intervals of about 2 months. Abscissa: $\alpha''\cos\delta$ or $15\,a^s\cos\delta$; ordinate δ''. Of course we are interested only in the small *variations* of α and δ.

Draw a smooth curve, showing the yearly path of the star.

2. In order to understand the causes of these displacements, we must in the first place eliminate the *precession*.

Simple but very practical precession tables are found in the *Star Atlas*. Compute what change this will produce in the coordinates of your star, at intervals of two months, and correct first of all the apparent positions for this effect.

3. Draw again the path of the star in the course of the year. You find a curve which is roughly circular. Determine approximately the radius.

4. Bradley hoped to discover the *parallax* of the star. Consider whether the path

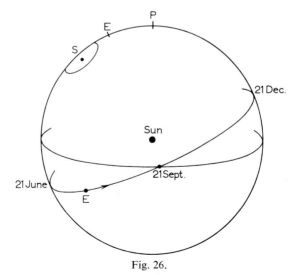

Fig. 26.

which we found might be explained by this effect. Assume very roughly that γ Dra is located near the pole E of the ecliptic ($\alpha = 18^h$, $\delta = 66°$); because of the parallax it would describe a small circle, corresponding to the circular motion of the earth (Figure 26). At what moments would the apparent declination be greatest? Does this correspond to the observations?

(Since then we know the parallax of many stars. Look in the table of exercise B24: if γ Dra is not found there, its parallax must be $<0''.2$. Draw this maximum angle on the scale of your diagram.)

5. Bradley had therefore to look for another explanation. The numerical value of the radius struck him, because it nearly coincided with the value, obtained by Rømer for the solar *aberration* effect and corresponding to the ratio: $V_{earth}/C = 20''.48$. He then assumed that the same phenomenon applies to the stars. For *that* hypothesis, at what moment should the declination be greatest? Is this confirmed?

6. We have now explained the apparent displacement by a combination of the precession and the aberration. There might be other effects.

(a) Look for the *proper motion* (Halley, 1718), which might be combined with the precession. In the table of Mean Places, find μ_α and μ_δ. To our approximation this may be neglected.

(b) When Bradley continued his observations of γ Dra during the following years, he found that there was another oscillation with a period of 18.7 years. In 1748 he explained this by the *nutation* of the earth's axis.

The star's position is shifted by an amount up to $\pm 9''.2$ in latitude, and up to $6''.8$ along the parallel of longitude. Looking to the star map, you see that for this star the shift in declination will be almost the same as the shift in latitude, being almost negligible in the course of one year. Negligible is also the nutation in α. The effects of nutation, however, become very apparent in the course of a 19 years period. This is nicely illustrated by a diagram on scale in DANJON (1952–53).

Plot also in your diagram the 'mean position' of γ Dra.

Probably it will be considerably different from the centre of its yearly path, as you would estimate this at first sight. Danjon's drawing explains this: the 'mean places' are positions, corrected for aberration and nutation; the mean position should thus not be compared to the path for one particular year, but to the path for 19 years.

(7.) Repeat the constructions for other stars.

TABULATION

Apparent position of γ Dra

	α 17^h 55^m	δ 51° 29′	prec. in α	prec. in δ	α-prec.$_\alpha$	δ-prec.$_\delta$
Jan. 1						
March 1						
July 1						
Sept. 1						
Nov. 1						
Dec. 31						

References

BECKER, F.: 1960, *Einführung in die Astronomie*, Mannheim, p. 45.
DANJON, A.: 1952–53, *Astronomie générale*, Paris, pp. 105–106.
SHAPLEY, H. and HOWARTH, H. E.: 1929, *Source-Book in Astronomy*, New York, p. 45, quotations
from Bradley.

Preparation

For each pair: *Astronomical Ephemeris*, any year.

For each student: rectangular coordinate paper; *Star Atlas*.

A17. THE MEASUREMENT OF ASTROGRAPHIC PLATES

1. How the Astrographic Catalogue has been Obtained

MATERIAL

You have received a sheet of the *Astrographic Map*, the celebrated 'Carte du Ciel', obtained by the collaboration of 18 observatories all over the world. This sheet is valuable and should be carefully handled. No pencil marks, no use of an eraser! If necessary, a sheet of transparent paper should be superimposed, on which special stars can be marked.

Three exposures have been successively made for slightly different positions of the plate; thus each star is represented by three tiny dots and no confusion with accidental specks is liable to occur. Moreover a system of réseau lines has been impressed upon the plate. For some *reference stars*, precise coordinates have been measured with the meridian circle. With respect to these, the coordinates of all the other stars may be determined, many of which are too faint for the meridian circle. We shall follow step by step the classical method by which the astrographic catalogues have been established.

The section of your catalogue, corresponding to the photograph, refers to a zone of 2° in declination, namely 1° on each side of the parallel δ_0. After an introduction, the plate descriptions are ordered in the succession of the coordinates α_0, corresponding to the centre of the plate. The coordinates α_0, δ_0 to which the telescope was pointed and which correspond to the centre of the reseau are found on top of the list. Only the brightest stars have been reduced (above 11^m). The reference stars are indicated by bold characters. The catalogue gives the rectangular coordinates x and y, measured on the original photograph and expressed in units of 2 mm, corresponding with about 1'. Identify a few stars; watch the direction of increasing x and y coordinates!*

At the end the reference stars are listed again and in the same succession. However, here are given the precise α and δ, determined with the meridian circle and reduced to the epoch 1900.0 (or 1950.0).

We shall see now how these coordinates are related to the measured distances x and y.

PROCEDURE (L)

1. Select three of the reference stars which are not too bright and rather far from each other, and list in a table the following data:

α; δ (for α the number of hours is not printed, since α never differs by more than 4^m from α_0). Precision: $0^s.1$ and $1''$.

$\alpha - \alpha_0$, $\delta - \delta_0$ (α_0 and δ_0 as mentioned on top of the photograph).

* In some catalogues the units correspond with 5'.

$(\alpha - \alpha_0) \cos \delta$ and $\delta - \delta_0$ in minutes of arc and 2 decimals; the factor $\cos \delta$ is slightly different from one star to the other (4 decimals). – These are approximately the coordinates which we may expect to measure on the plate. We call them: the *approximate coordinates*.

2. However, since a portion of the celestial *sphere* has been imaged on the *plane* of the plate, a small but rather complicated correction has to be applied which we shall call the 'flattening correction'. Provisionally it will be neglected.

We have now obtained the *standard coordinates* of the stars (= tangential coordinates), defined as follows.

Assume that the centre of the réseau corresponds exactly to the position α_0, δ_0; that the lines are exactly parallel to the directions of α and δ; that the plate was perfectly perpendicular to the optical axis; that there was neither refraction nor aberration. In this ideal case of a pure central projection on the tangent plane the rectangular coordinates of the star on the plate, expressed in terms of the focal length, will be called the standard coordinates ξ and η.

3. The assumptions, mentioned above, are of course never strictly fulfilled, moreover the scale does not quite correspond to the relation: $1' = 2$ mm. The measured coordinates x, y will therefore differ slightly from the standard coordinates ξ, η. It can be shown that all these deviations can be expressed by linear equations, and so their sum:

$$\xi - x = a + bx + cy$$
$$\eta - y = d + ex + fy.$$

Each star gives such a pair of equations; the constants a, b, c... are the same for the whole plate. (In our case the deformation by reproduction and by contraction of the paper will be included in these equations.)

The measurement of x and y has been made with a microscope, moved by the carefully calibrated screw of a plate-measuring machine. We shall imitate this operation with a simple millimeter scale, which we read to 0.1 mm (use a magnifier!). The position of the star which we measure is the centre of gravity of the three dots.* Tabulate the values of x and y, the *measured coordinates*.

Write down the three equations for $\xi - x$ and so for $\eta - y$ ($\xi - x$ and $\eta - y$ in two decimals). Solve for a, b, ..., f by elementary computations; use 4 decimal logarithms.

It is interesting to notice how laborious such a simple operation is; errors may be avoided if each of the two students is working in a strictly independent way and if they check their results at each step.

Examine critically the values of the constants. Centering errors appear mainly in a and d; the scale influences b and f; errors of orientation appear in c and e. In order to estimate the greatest shifts, due to b, c, e, f, substitute $x = 100$ mm or $y = 100$ mm.

COORDINATES OF AN UNKNOWN STAR

4. Reversely, we are now able to derive α and δ of an unknown star from its measured

---

* For some zones the three images have been given very different exposures; in that case we take the middle point of the strongest two dots.

coordinates x and y. As such we select a fourth reference star, which is not so bright that its position would be uncertain. We assume that its α and δ are unknown and we measure x and y (expressed in units of 2 mm). From our two equations, of which the constants now have been ascertained, compute ξ and η, and finally α and δ.

Compare with the coordinates, given in the list of reference stars. A perfect agreement cannot be expected:

(a) because of measuring errors; (b) because we have neglected the flattening correction; (c) because our plate constants have been derived from 3 stars only.

Note that by this method the coordinates are automatically reduced to the epoch 1900.0, even if the plate has been taken at another moment. Only the proper motions of the stars over a few years have been neglected.

The procedure followed may be summarized by the following flow diagram:

equatorial coordinates α, δ		approximate coordinates		standard coordinates ξ, η		measured coordinates x, y
			↑ flattening correction		↑ deformation correction	

2. The Practical Use of the Astrographic Catalogue

We shall now see how well the Catalogue has been arranged for practical use, in all cases where the coordinates α, δ of a faint star have to be found. The instructions, given in the introduction of each volume, are slightly different for the several observatories. Also the tables of corrections differ from one volume to the other, depending on the part of the sky to which it refers.

Often the plate constants a and e are already included in the values for the central α and δ. In other cases provisional scale corrections are applied before the precise b, f are computed.

PROCEDURE

1. Select one of the reference stars, which will play the role of 'unknown star'; it may be the same star, studied at the end of part 1 of this exercise. Now we take over its measured coordinates x, y, as given in the catalogue with their full precision.

Look up the introduction and see whether a numerical example is given; try to follow this closely. Realize for each operation how it fits into the general scheme.

2. Then apply a similar procedure to the star which you selected. From the example you may see the number of decimals required. For the correction for deformation you find the plate constants at the top of the plate description.

For the small flattening correction, accept the formulae given in the introduction and see how they have been reduced to tables, by which the computations are considerably simplified.

Finally find α, δ and compare with the values, found with the meridian circle.

References

SMART, W. M.: *Spherical Astronomy*, chapter 12.

TURNER, H. H.: 1912, *The Great Star Map*, London.

For short methods, to be applied if we are only interested in one star, see A. König: 1960, in *Stars and Stellar Systems* (ed. by G. P. Kuiper and B. Middlehurst), Vol. II, Chicago, p. 483.

Preparation

For each pair:

For part 1, a sheet from the *Astrographic Map* with the corresponding *Catalogue*. Very suitable are the zones of *Bordeaux* (6 volumes). In those of S. Fernando we find the directly measured x, y, as well as the same coordinates corrected for flattening; the flattening correction is obtained here by a huge table in vol. I.

For part 2 of this exercise we recommend: *Bordeaux* (6 vol.), *Uccle-Paris* (2 volumes).

Magnifier; millimeter scale; table of logarithms (5 decimals; for part 2, 6 decimals).

A sheet of transparent paper.

Demonstration of a plate-measuring machine.

Note. If time must be saved, one pair of students may derive the constants for $\xi - x$ and the value of α, while the other pair computes the constants for $\eta - y$ and the value of δ.

A18. TO GRIND A TELESCOPE MIRROR

(Three evenings)

This exercise is not intended to make a perfect piece of optics, but just to show the principle of the methods used in practice and to convince our students that amateur telescope making is surprisingly simple and accessible. Further technical details may be found in many books on amateur telescope making.

The Principle

If a circular glass plate is moved to and fro over a similar, fixed glass plate – some grinding powder being introduced between the plates – the lower will of itself become convex and the upper will become concave. Let us call them: *the template T* and *the mirror M*.

We use ordinary plate glass, at least 1 cm thick, out of which discs have been cut each with a diameter of 11 cm; the rims should be blunt.

Grinding will be done on an empty upturned wooden barrel, around which we will have to be able to move freely. The template is fastened in the centre of the upper end by means of three little wooden blocks and a wedge (lower than the surface of the glass disc! Figure 27 and 27a).

(a) A *stroke* is a forward and backward motion, by which the centre of the mirror moves along the diameter of the template. The longer the stroke, the quicker the mirror deepens.

(b) At intervals of one minute or so you take one step around the barrel, always in the same sence. This is a precaution necessary in order to ensure that the shape of the mirror becomes perfectly concentric.

Procedure (L)

1. *Coarse grinding* (half an hour). We start grinding with carborundum nr. 180 using long strokes, the centre of the mirror moving over almost the entire diameter of the

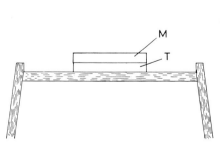

Fig. 27.

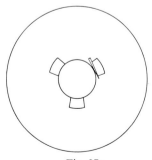

Fig. 27a.

template. Sprinkle two level tea-spoons of carborundum on the template and add some water (with a drop-bottle), till a rather fluid paste is formed. Place your hands gently on the mirror and make long strokes for a few minutes, till you can hear that the carborundum is no longer effective. Then add a new amount of carborundum, not taking the trouble to remove the old paste, and go on grinding. Rather strong pressure should be applied.

After about 15 min the mirror is rinsed under the water tap. While the surface is still wet, you may determine the radius of curvature: stand at least two meters away from the mirror and move a flashlight across the axis; looking at the mirror you will see directly whether the light is before or behind the centre of curvature.

When the mirror takes a better shape, you hold it at such a distance that an image of the lamp is formed on a screen, next to the lamp. The lamp is then at the centre of curvature, the radius of curvature is twice the focal distance, which in our case should be about 1.50 m. Go on grinding until this is attained.

2. *Fine Grinding* (two hours). Wash the mirror and the template carefully under the water tap, especially the edges and rims. The barrel should also be cleaned (the water may run into the barrel through a hole). Any particle of the coarse carborundum which may be left, might spoil the subsequent work. The same applies to the next steps of fine grinding.

We now apply short strokes, not intended to deepen the mirror but to make it more perfectly spherical. The centre of the mirror should move up and down over $\frac{1}{3}$ only of the diameter of the template. The pressure should again be rather strong.

Successively use: carborundum F   for 15 min;
                              FF   for 15 min;
                              FFF   for 15 min.

In between each operation, wash the mirror and remove all dust of the preceding step. The surface becomes fine-grained and smooth.

Finally apply carborundum 800. When the typical sound of grinding has disappeared, go on working for at least 30 min, without adding more powder. The pulverized crystals of carborundum become finer and finer and so does the grain of the glass surface.

Check again the radius of curvature.

3. *Polishing* (1 to 2 hours). Wear a dust-coat! In the interval between the two evenings of this exercise, your template has been covered with a layer of pitch.

Place the mirror and the template in lukewarm water (50°). After some minutes dry the template and then moisten it with some turpentine, this is to make the pitch adhere better; then wind a rim of sellotape around it, so that a shallow dish is formed.

The pitch has been prepared by melting together 1 part of pitch and 2 parts of resin. A 3 mm thick layer of that mixture is poured on the template; allow it to cool off somewhat. Now take the mirror out of the water and press it against the pitch layer, rotating it slightly. Remove the sellotape collar and with some pressure, slide the mirror to one side. With a moistened knife quickly cut parallel grooves into the pitch, first lightly, then more deeply, about 2 cm apart, asymmetrically to the centre, so that the circular area is subdivided in squares.

The grooves should be cut into a V-shape and should reach the glass. Press the wet mirror against the template again and make a few short strokes, till all facets are in contact with the mirror. Half an hour before the polishing starts, sprinkle some polishing powder on the mirror and put it on the template, with a heavy book on top, to make sure that the pitch surface takes the form of the mirror.

Polish with short strokes, with 'rouge' or (by preference) with cerium oxide. These powders should be kept free of any dust! – Always keep the mirror very carefully in contact with the pitch over its whole surface. After one hour, wash the mirror, dry with cotton wool, and inspect the surface. Go on polishing for two half-hour periods. Or longer if you have the patience!

The mirror will not be really smooth, but sufficiently so in order to proceed to the following operations.

When the mirror is dry, a handle *H* is luted to the back with a suitable cement. In order to avoid the cement coming into contact with the silvering solution, apply, with a brush, a protecting rim *P* of molten paraffin around the handle.

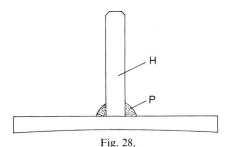

Fig. 28.

4. *Silvering* (Brashear's recipe). Again wear a dust-coat. The *reducing solution* has been mixed a few weeks in advance:

Water 1000 cm³, white sugar 100 gram, alcohol (94%) 125 cm³, concentrated nitric acid 4 cm³. In the course of time the solution is inverted to glucose, which is the reducing agent.

The *silvering solution* may be prepared for a number of students at the same time.

Prepare: *A*. Distilled water 1000 cm³, silver nitrate 100 g;
             B. Distilled water 750 cm³, potassium hydroxide 50 g;
             C. Distilled water 1000 cm³, silver nitrate 15 g.
The following measures are given for *one* mirror.

1. Take 38 cm³ of *A*.

2. Add slowly 2 cm³ of concentrated ammonia, stirring all the time: a precipitate is formed, which should *just* be dissolved by adding cautiously one or two cm³ more; *one* drop too much removes the last turbidity and makes the solution too clear.

3. Add 35 cm³ of *B*, stirring all the time. The liquid becomes dark brown.

4. Add again ammonia, *just* enough to clear the liquid.

5. Add slowly solution *C*, stirring, till the liquid becomes slightly turbid (colloidal silver). Filter through cotton wool and pour the liquid into a low cylindrical glass; the temperature should be between 18° and 23 °C. The solution thus made should be used without undue delay and in no case be left till the next day; there might be formation of silver fulminate and an explosion!

The mirror should now be prepared for silvering. The surface is cleaned with cotton wool, dipped in nitric acid (use rubber finger protectors! avoid splashes!);

then washed carefully under the tap, and finally rinsed with distilled water (temperature 20°–25 °C).

You have poured 30 cm³ of the silvering solution in the cylindric vessel. Now add, quickly, 10 cm³ of the reducing solution, and dip your mirror in the mixture; avoid air bubbles by first inclining the mirror; then using the mirror to stir with, but avoid touching the bottom of the glass vessel. After one or two minutes the liquid becomes dark brown, and after about 5 minutes minute scales of silver are floating on the surface and no further silver deposit is formed.

Take the glass vessel to the sink, wash the mirror with running water, never touching the surface; rinse with distilled water. Dry your mirror in vertical position, by preference in the air current of a fan. – If time has to be considered, you may now proceed directly to the Foucault test. The silver layer is not yet polished, but this hardly matters and may be done later very quickly.

5. *To Polish the Silver Layer.* (Dust coat or overall! Stains of polishing rouge cannot be removed from clothing.) The day after the silvering, or later, we polish the mirror with chamois-leather and the finest rouge. The blue haze on our mirror disappears within a few minutes, but microscopic scratches can not be avoided.

You can now determine beautifully the radius of curvature by finding out at what distance the image of a small electric lamp coincides with the lamp itself.

6. *The Foucault Test.* We want to test whether all rays, emitted by a point-source in the centre-of-curvature and reflected by the mirror, converge exactly at one and the same point.

The room should be only dimly lighted. We need a small table, at which we sit and prepare the observation. The source of light is a small incandescent lamp *L*, around which slides a well-fitting metal cylinder, provided with a tiny hole (0.5 mm). Let this cylinder first be removed.

At the other end of the room, on another table, the mirror *M* is put in the right position by clamping the handle into a holder. It should throw the reflected image of the little lamp quite near to the lamp itself and at nearly the same height (use a piece of paper as a screen). Once the image *C* has been localized, push the metal cylinder over the lamp, taking care that the transmitted beam of light fully illuminates your mirror (check this!).

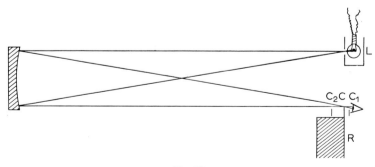

Fig. 29.

Take a little wooden block, to which a razor blade R has been fastened; on the blade we have glued a piece of white paper, leaving only the edge free. Catch the image of the tiny hole on the white paper, then on the razor blade. If you move the block aside and put your eye at the place of that luminous point, you will see the mirror full of light. While you are looking, the test is made by moving the razor's edge slowly *from left to right* across the beam. This should be done precisely at the centre of curvature C; do it first at a point C_1, 2 cm nearer to you: you will see a shadow moving over your mirror *from right to left*; next do it at a point C_2, 2 cm beyond the centre of curvature: the shadow moves *from left to right* over the mirror. You easily find a point in between, where neither the first, nor the second case applies. You have now approximated as well as possible the centre of curvature, but you notice that some parts of the mirror show a shadow, while other parts are bright. This is the critical point! – In most cases the mirror looks as indicated in the Figure.

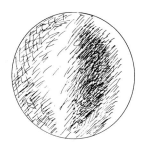

This shadow pattern suggests a curved surface, the relief of which is correlated with the shape of the mirror surface itself. This may be improved by further polishing. – But as this is too time-consuming, we prefer to enjoy the sensitivity of the set-up by a few simple experiments:

(a) Take your handkerchief from your pocket and put it under the mirror: you see the hot air rising like streaks of smoke.

(b) Put your finger on the back of the mirror for some seconds; there is a local heating and dilatation of the glass, which you clearly perceive in the Foucault pattern.

(c) Realize how the rays of light are converging and how the shadow pattern originates.

References

Amateur Telescope Making, Munn and Co., several editions.
SIDGWICK, J. B. (quoted on p. XIV), ch. 7.
THOMPSON, A. J.: 1947, *Making your own Telescope*, Cambridge, Mass.
WENSKE, K.: 1946, *Spiegeloptik*, Mannheim.

The methods of grinding, polishing and testing, used for big telescope mirrors, are in principle the same; only the strokes are made by automatic machines. The Foucault test may be refined to a quantitative test and it is followed by another test, the Hartmann-test.

Preparation

A. For each pair: two circular glass plates, 11 cm diameter, at least 1 cm thick; one of these is luted to a handle by a suitable cement, the other is fixed at the centre of an upright barrel.

A flashlight; a drop-bottle; carborundum 180; F; FF; FFF; 800.

B. Each of the templates is covered by a layer of pitch, in which grooves have been cut.

Polishing powder (*cerium oxide* by preference); molten paraffin; brush; reducing solution (see text) prepared several weeks ahead; silvering solution *A, B, C* (see text); two graduated glasses of 50 cm^3; nitric acid, fairly concentrated; cotton wool; rubber finger protectors; 2 cylindrical glass vessels for cleaning and silvering the mirror; distilled water.

For the whole group: an electric fan.

C. Soft chamois leather; finest polishing rouge.

Two set-ups for Foucault test: mirror holder, incandescent lamp, perforated cylinder, wooden block with razor blade and piece of white paper.

The test is made for each of the mirrors in succession, every time one of them is considered to be finished.

A19. OPTICS OF A SMALL TELESCOPE

For many of our observations we shall need a small telescope. Our instrument is a refractor, with a 4 cm achromatic objective and a focal distance of about 50 cm; consequently the aperture is $f/12,5$.

If you wear glasses, remove them when observing through a telescope, unless you are astigmatic or unless your eyes require a strong correction.

Procedure (*Mainly* L.)

1. First study the way in which your telescope moves. Is it an altazimuth or a parallactic mounting? How can it be clamped? The clamping screws should be just turned gently, never tightened with effort!

When looking through the telescope, always keep the eyes at rest, avoid straining.

2. Direct the telescope at a distant object: (a) looking through a window pane; (b) through an open window; (c) hold a piece of window-glass before the objective.

Are you able to observe near objects also? How near? What determines this minimum distance?

3. The eye-piece can be taken off. Is it a Ramsden or a Huygens eye-piece? In the first case you can use it as a magnifier, it gives a real image of a lamp. In the second case this does not work.

Toward which side are the convex sides of the lenses directed in each of the two cases?

Measure the distance between the front of the objective and the back of the eye-piece, focussed at infinity. Then unscrew cautiously the whole ocular end of the telescope.

Where are the cross-wires? Take care, don't touch them! – Where is the diaphragm? Determine the precise place where the objective forms an image of distant objects (roofs of houses, street lights).

Take the eye-piece in hand and measure where an object must be placed with respect to it in order to be seen clearly. Apparently this is the place where the image, formed by the objective, is located when the telescope is assembled and focussed at infinity. – Make a drawing *to scale*, showing the objective, the image of a distant object and the eye-piece in their relative positions.

4. Determine the focal distance of the eye-piece as a whole. It is not a single lens, but a lens system, and your method must take this into account. Take as an object two distant lamps, the line joining them being perpendicular to the line of sight B and having a length V. Find their images, separated by a distance v.

From the relation $V/B = v/b$ you are able to deduce $b \simeq f$. What is now the magnification of your telescope?

5. Focus the telescope at infinity. Now direct it towards a well-lighted wall and notice the *eye-ring*. Hold your eye behind or before the eye-ring and notice how the field changes.

Ascertain the origin of the eye-ring by putting a pencil on the objective and looking for its image; you see now that the eye-ring is an image of the objective.

Measure the diameter of the ring with a millimeter scale (in a stand) and a magnifier. Again find the magnification of your telescope.

6. The brightness of a point-source (star), observed through a telescope is much greater than the brightness observed by the naked eye, namely, in the proportion of the area of the objective to the area of the pupil (diameter at night = 8 mm). This applies only if the eye-ring is wholly admitted through the pupil. What is the gain in the case of your telescope?

7. As soon as an object is not *very* far (= 'infinitely far'), you will be surprised how quickly you notice this in the position of the eye-piece.

First focus accurately on an object at least 1 km distant; then choose nearer objects at a distance V, carefully adjusting the eye-piece every time and measuring its position. With respect to the position for infinity the shift Δ will be found proportional to $1/V$. For an object at x times the focal distance, Δ will be found approximately equal to f/x. – Derive this useful formula!

8. First see whether the cross-wires are in focus. Then focus the telescope as a whole on a far-off object. Check whether the cross-wires have been adjusted at precisely the right place: is there any parallax? Shift the position of the eye-piece slightly; what is the accuracy of your focussing? – If you twist the ocular around the optical axis of the telescope, the object which coincides with the intersection of the cross-wires should remain precisely the same.

9. When looking at a bright field, you observe the image of many tiny dust-specks. Where are these? Rotate the eye-piece and notice whether they are moving too. If possible, rotate separately each lens of the eye-piece.

Find the angular extent 2φ of the field, from the diaphragm aperture and the focal distance. [A direct determination is found in exercise A5.]

How does the field appear if you use the eye-piece without its front lens ('field lens')?

10. Look at any distant landscape and cover part of the objective with a piece of paper. How does the field change? And the brightness?

11. In order to observe the effects of diffraction in a telescope we shall make them many times stronger than normal. Diminish the objective aperture to 1 mm by means of a diaphragm. Now look at a distant, very small but bright source of light (incandescent lamp): observe the diffraction pattern; make a sketch, representing the intensity distribution across a diameter.

12. For the circular diaphragm, substitute a piece of wire gauze or any other screen and observe the diffraction. Imitate an objective grating by putting a comb before the objective.

Put a colour filter between the eye-piece and your eye. Look at the diffraction patterns in red light and in blue light.

References

DANJON, A. et COUDER, A.: 1935, *Lunettes et telescopes*, Paris.
JOHNSON, B. K.: 1960, *Optics and Optical Instruments*, New York.
ROTH, G. D.: 1960, *Handbuch für Sternfreunde*, Berlin.
SIDGWICK, J. B.: 1956, *Amateur Astronomer's Handbook*, London.

Preparation

For each pair: experimental telescope; measuring scale and stand; a piece of window-glass; diaphragm 1 mm; wire gauze; red filter and blue filter.

For all: two lamps, a few meters away, and about one meter apart from each other; an automobile lamp with a very short filament, at a distance of some 30 to 50 meters.

THE MOTIONS OF CELESTIAL BODIES

A20.   THE ORBIT OF THE MOON

The Problem

To determine by simple means the apparent orbit of the moon.

The position of the moon is determined on one evening. Further determinations are made on other evenings, at home or at the observatory. Finally all data are assembled and combined.

Procedure

We shall measure the angular distance of the moon from the neighbouring stars with the *cross-staff*, a simple instrument already used by sailors in the time of Columbus and used by fishermen up to the beginning of the 19th century (Figure 5).

The extremity A of the rod is pressed against your cheek-bone, just under your eye; the cross-bar is moved to such a distance, that the two points B and C appear to coincide with the objects between which the distance has to be measured. Gently

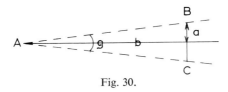

Fig. 30.

clamp the cross-bar with the screw, correct the adjustment slightly. The angle $BAC = \vartheta$ is then read directly on the graduation I (the small distance between A and the pupil has been taken into account):

$$\tan \frac{\vartheta}{2} = \frac{a}{b}.$$

For smaller angles, use the short side of the cross-bar (width $= 5$ cm), to which corresponds another scale II on the rod. For the smallest angles use the still narrower end (2.5 cm) and divide by two all numbers of the scale II.

When making measurements at night, it is necessary to illuminate the cross-bar with a flash light: it stands out beautifully against the dark night-sky.

1S. Measure the distance between some distinct landmarks (in $0°.1$). Each distance is measured 3 times and averages are taken.

2S. Measure some angular distances between stars:

        from Deneb to Vega (compare with A2)

        from α Cyg to β Cyg

        from Capella to Aldebaran

        from α UMa to the Pole Star

3S. Measure the distance of the moon accurately from 3 neighbouring stars (not farther than 30°). Since your distances will have been measured from the limb of the moon, add 0°.2 to each distance.

Now take a map of the ecliptical constellations and, with compasses, draw the arcs corresponding to the measured distances. The centre of the moon is located at the intersection.

4. Such measurements should be repeated as many days as possible during a lunation. The positions determined are plotted in the ecliptical map and connected with a smooth line. Always note also the time of the day. Measurements of all students of the group are combined. – Small 'irregularities' in your plot may be due to the effects of parallax, which are by no means unimportant.

5L. From this, derive:

the duration of the sidereal month;
the longitude of the ascending and descending node;
the inclination of the orbit with respect to the ecliptic.

Compare to the 'official' values.

6L. Because the moon's orbit is elliptic, the apparent diameter changes appreciably in the course of a month.

Take a prismatic spectacle glass with a refracting angle α of one degree (nominally). The deviation is approximately: $\delta = \alpha(n-1) = 0.53\alpha \simeq 0°.5$. Because for such commercial glasses the angle varies slightly from one side of the glass to the other, a piece of dark paper has been glued to it, leaving only a small part free.

Point your experimental telescope to the moon and insert the prismatic glass before part of the objective. You observe now a second image of the moon, shifted over about its diameter. It does not matter before what part of the objective the prism is inserted; the secondary image is always shifted towards the thickest side of the prism, in the direction indicated by an arrow on the paper diaphragm. Turn the prism into such a position that the shift occurs parallel to that lunar diameter which corresponds to the extremities of the terminator and which is seen in its full extent.

Estimate the overlap of the interval between the two images as a fraction of the lunar diameter. Then consult in the *Astronomical Ephemeris* the table on the motion of the moon, and note the angular diameter of the moon at this moment. – The observation should be repeated on one or two other days, when you may expect the angular diameter to have changed considerably. Avoid slender crescents and do not measure when the altitude is less than 10°.

The same simple device may be used to compare the angular size of the moon, low above the horizon and high in the sky. You will then be convinced that psychological effects may be very deceptive! At altitudes below 5° you will observe the refraction.

(7S.) Since measurements with the cross-staff are so quick and so easy, one may use this opportunity in order to test the precision of our measurements and to find the error curve. The measurement described under (1) is repeated a number of times, say 40 (20 times would be hardly sufficient), and statistics of the readings are made.

All measurements should be obtained by the same observer. Between successive measurements the cross-bar should be shifted, so that no remembrance of the previous position is left and that the observations are truly independent. Draw the error curve. Determine the mean error; you will find a value between $0°.1$ and $0°.2$.

The same statistics may be made at night by measuring the distance between two bright stars. You will find about the same mean error; but according to our experience the error curve is less regular than for day-time measurements.

(8S). If you have a sextant available, why not use it and follow how the diameter of the moon varies from day to day? The measurement is described in the exercise A13, section 5; only in *this* case we have to rotate the plane of the sextant very carefully around the direction of vision till the terminator of the direct and that of the reflected image are precisely in line. Reject the three first measurements, which are necessary in order to gain experience. Then take three readings r with the reflected image below the direct image, and three readings r' in the reverse position. Average and take $\frac{1}{2}(r - r')$.

This should be repeated on a number of days and compared to the *Ephemeris* data. – Avoid slender crescents and do not measure when the altitude is less than $15°$. It is a fine exercise and the results are satisfactory.

Preparation

For each pair: cross-staff (with double graduation); compasses; *Ecliptical Star Map* (to be purchased from: Sky Publishing Corporation, 49-50-51 Bay State Road, Cambridge, Mass. – Or from: Lehrmittel-Reinecke, Markkleeberg).

A few spectacle lenses, prismatic, refracting angle $1°$. The optical quality of such glasses is not always satisfactory, often they converge or diverge the rays of light; they should be selected by putting them before the objective of our experimental telescope and judging the quality of the image. Dark paper is glued on a great part of the prism, leaving free only $1\ cm^2$ of the best part. – The observations on the apparent diameter of the moon should be made by preference when first and last quarter coincide approximately with apogee and perigee.

The use of the cross-staff for the elementary determination of the moon's position was initiated by Stetson and also applied by Dustheimer, Shaw and Boothroyd.

A21. THE ORBIT OF THE MOON FROM EPHEMERIS DATA

If numerous direct observations of the lunar positions are not possible for practical reasons, or if we wish to study the lunar orbit over a longer interval of time (say: one year), we may use the tables of the *Astronomical Ephemeris*, which have been calculated on the basis of observed quantities and we may accept them as 'observed'. They are given directly in ecliptical coordinates, as if seen from the centre of the earth.

Procedure (L)

1. From the lunar tables of the Ephemeris, calculate:

the duration of the sidereal month (devise your own method!);
the inclination of the orbit with respect to the ecliptic.

2. Note the longitude of the ascending node at the successive passages through the node. A simple interpolation gives a precision of 1° or 0°.5, which is sufficient.

Plot these longitudes as a function of the number of the passage. Notice the regression of the nodes and estimate the lapse of time necessary for a complete revolution.

The regression of the nodes is not uniform. It disappears at two moments of the year: check that these are the moments when the sun's position coincides with one of the nodes (Table Sun of the *Astronomical Ephemeris*); at these moments the 'orthogonal component' of the solar perturbing force disappears.

3. The *Astronomical Ephemeris* gives also the angular lunar diameter. Note the longitude of the moon at those moments when the diameter reaches its maximum value (the moon at perigee). Notice how the longitude of the perigee varies in the course of a year: the perigee alternately advances and regresses, but the advance exceeds the regression and the net result is an advance. This results from the combined action of the normal and the tangential component of the disturbing solar attraction.

Reference

MOULTON, F. R.: *An Introduction to Celestial Mechanics*, New York, several editions, reprinted 1962, Dover Editions.

A22. LUNAR ECLIPSES

The Problem

Find from the *Ephemeris* when the next lunar eclipse, visible from our city, will take place. Or select another eclipse in the year for which you have the *Ephemeris*. We want to ascertain the circumstances of this eclipse by means of the data in the *Astronomical Ephemeris* and by a graphical procedure.

Procedure (L)

1. Look up in the *Astronomical Ephemeris* the right ascension of the sun and of the moon around the eclipse day; determine the UT moment of the opposition in right ascension (precision: 1 minute).

2. Take as the origin of your graph the centre O of the earth's shadow. We imagine our drawing moving together with this shadow, so that O remains in its centre. On the vertical axis of declinations will be found the centre of the moon at the moment of opposition. Our graph must show how the moon gradually approaches the shadow of the earth and then recedes from it.

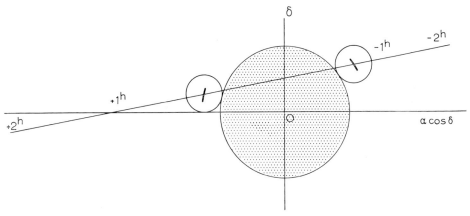

Fig. 31.

3. Look up in the *Astronomical Ephemeris* how the right ascension of the moon and that of the sun vary between two hours before and two hours after the opposition. Take the differences α (moon) – α (earth shadow), reduce to minutes of arc (1 minute of time corresponds to $15' \times \cos\delta$ minutes of arc).

4. Do the same with the declinations. Bear in mind that the declination of the earth's shadow *increases* when that of the sun *decreases*; and the other way round.

5. Now plot the positions of the moon's centre with respect to our origin of coordinates, namely, for -2^h, -1^h, 0^h, $+1^h$ and $+2^h$ after the opposition. Take 1 mm = 1′. Choose the correct sense for the $\Delta\alpha$-axis!

6. Draw the path of the moon in our coordinate system. Is it a straight line?

7. The angular radius of the earth's shadow is equal to:

horizontal parallax of the moon + parallax of the sun −
solar radius (in angular measure).

The necessary data are found in the *Astronomical Ephemeris*. Draw a circle around the origin, having the correct size.

8. Also draw short arcs, having a radius equal to the radius of the earth's shadow + the radius of the moon. Their intersection with the path of the moon marks the position of the moon at first and at last contact. Draw the lunar circumference at these two moments.

Similarly, draw the position of the moon at the beginning and at the end of totality.

(9.) Read the position of the moon's centre for the 4 moments of contact. Calculate when these occur (in UT).

Also measure with a protractor the position angles of the contact points along the lunar circumference.

Make a tabular summary and compare with the data of an almanac.

(10.) What do you find about this eclipse in VON OPPOLZER's *Canon der Finsternisse*?

References

LINK, F.: 1956, *Die Mondfinsternisse*, Leipzig.
SMART, W. M.: *Textbook on Spherical Astronomy*, Cambridge (chapter XV).
VON OPPOLZER, TH.: 1887, *Canon der Finsternisse*, Wien. – English Translation: 1962, New York.

Preparation

For each student: a sheet of rectangular coordinate paper.

For each pair: *Astronomical Ephemeris*; compasses.

A23. THE POSITION OF THE PLANETS IN THEIR ORBITS

The Problem

To draw approximately the planetary orbits and the momentary position of planets, in particular for Mars.

The orbits of some of the planets are ellipses, which can be approximated by circles, of which the centre is located at a small distance from the sun ('curtate orbits'). Since the inclination to the ecliptic is small, we may draw them in the plane of the ecliptic.

We shall make use of the *elements* of the planets, which are found in any textbook on astronomy, in the *Astronomical Ephemeris* and *Norton's Star Atlas*.

Procedure (L)

1. Let us take the planet Mars as an example.

The *sun* is represented by a point in the centre of our sheet of paper. From there, draw a horizontal line towards the *Vernal Equinox*. Read from your table of elements the longitude Ω of the *ascending node* of Mars, and draw the line of nodes.

2. Similarly, draw the line towards the *perihelion* of Mars, paying attention to the way in which the π, 'longitude of the perihelion' is defined.

3. Take as a scale: 5 cm = 1 astronomical unit. The distance from the centre of an ellipse to its focus is: $c = ea$ (e = eccentricity, a = major axis). Take e and a from the table of elements, reduce c and a to the scale of the drawing. Calculate also the minor axis b.

Check whether the representation of the ellipse by an eccentric circle is sufficiently accurate (apply the approximations allowed for small quantities).

$$b = a\sqrt{1 - e^2} \simeq a(1 - \tfrac{1}{2}e^2).$$

4. Draw the orbit of Mars. That part which is above the ecliptic should be a continuous line; the remaining part should be a broken line.

5. In the same way draw the orbits of the earth and of Eros.

6. What is the geocentric longitude of the sun on March 21? What is the heliocentric longitude of the earth on March 21?

7. The position of a planet in its orbit is determined by the angle perihelion-sun-planet, called the *true anomaly* ω, which is little different from the *mean anomaly M* (that which the planet would have if, starting from perihelion, it had the same angular velocity all the time).

Knowing the epoch of the latest passage of Mars through its perihelion, compute its mean anomaly to-day (use the tables of Julian days!). From elementary planetary theory:

$$\omega = M + 2e \sin M + \cdots.$$

Draw the radius vector towards the planet and plot the position of Mars. What is its heliocentric longitude?

(8.) Where is the earth at this same date?

(9.) Measure in your drawing the geocentric longitude of Mars and find in *Norton's Star Atlas* the constellation where Mars may be observed.

(10.) The *heliocentric latitude* is approximately given by: $\beta = i \sin(\lambda - \Omega)$. The *geocentric latitude* is found from the consideration that the planet, at a distance r from the sun, is located at a height βr above (or below) the ecliptic plane. If A is the distance from the earth to the planet, the geocentric latitude will be: $\beta(r/A)$.

Measure r/A on your map, compute the geocentric latitude and compare with the ecliptic map.

(11.) What is the angle between our line-of-sight and the boundary plane between light and dark on Mars (the 'terminator plane')? Draw the disc of the planet as it should be visible to-day.

(12.) When Mars is in opposition, its distance to the earth is relatively small. In which part of the year will the next opposition occur? In what constellation will Mars then be?

Reference

Graphs indicating the positions of the planets during the current year, to be found in astronomical almanacs or yearbooks.

Preparation

For each student: rectangular coordinate paper; big protractor.

For each pair: table with the elements of the planets, or *Astronomical Ephemeris*; compasses.

A24. THE ORBIT OF MARS, AS DETERMINED BY KEPLER

The Problem

In this exercise we shall repeat, in condensed and simplified form, the celebrated investigation by which Kepler determined the orbit of Mars and the laws of its motion. We wish to save time and to make simple drawings to scale, instead of solving triangles by trigonometry.

While Kepler worked within one minute of arc, we shall be content with a precision of $1°$. Great care in drawing is required, in order to reach even this modest accuracy.

Data

Basic for Kepler were Tycho's tables, giving the position of the sun between the stars during the year; this means that *the heliocentric longitude of the earth* is known as a function of time.

Next come the observations of the angle $\gamma =$ sun–earth–Mars. Especially important are the moments of opposition, when $\gamma = 180°$, because at these moments, and only then, we know directly *the heliocentric longitude of Mars*. By a kind of interpolation process this longitude may be found for any moment of time.

TABLE I
Positions of Mars

Day	Julian Day	Radius vector of Earth	Heliocentric longitude of Earth ($\lambda_\oplus$)	of Mars (λ_δ)	Angle $\gamma = SEM$
21–4–1920	2422–436	1.005	211°	211°	180°
9–3–1922	3–123	0.993	168°	211°	98°
11–6–1924	3–216	1.017	259°	259°	180°
25–1–1924	3–810	0.985	124°	211°	60°
24–8–1924	4–021	1.011	330°	330°	180°
11–7–1926	4–708	1.016	288°	330°	91°
4–11–1926	4–824	0.992	41°	41°	180°
28–5–1928	5–395	1.014	247°	330°	59°
21–9–1928	5–511	1.005	358°	41°	93°5
22–12–1928	5–602	0.984	89°	89°	180°
9–8–1930	6–198	1.014	316°	41°	58°5
28–1–1931	6–369	0.985	127°	127°	180°
14–12–1932	7–055	0.984	84°	127°	101°
2–3–1933	7–133	0.991	161°	161°	180°
1–11–1934	7–742	0.992	38°	127°	59°5
7–4–1935	7–899	0.978	196°	196°	180°
20–5–1937	8–673	1.012	238°	238°	180°
23–7–1939	9–468	1.016	300°	300°	180°

From the observations of Tycho, David Fabricius, and Kepler himself, the oppositions were accurately known over a period of 24 years. For a number of conveniently
selected days in recent years, the most important observed coordinates are summarized
in Table I.

Procedure (L)

1. Determine the synodic period of Mars. This is the mean interval between two
consecutive oppositions. Look at the Julian Days and notice that these intervals are
not all the same (why?). We might take an average over an interval of a number of
years. But even then a precise result will be obtained only if the first and the last
opposition of our interval occurred in almost the same part of the Martian orbit.
Find a suitable combination; or, still better, take the mean between two suitable
combinations and derive the synodic period. (Always count in Julian days!)

2. From this, calculate the sidereal period. (It should be $687^d.0$.)

3. By a first triangulation, which will not be repeated here, Kepler had determined
the precise shape of the earth's orbit. On a sheet of paper draw a circle with a radius
of 5 cm. The eccentricity of the earth's orbit amounts to $e = 0.0167$ and should be
taken into account. The length of the perihelion is $\omega = 101°$. Also draw the radius
vector towards the Vernal Point: this will be a horizontal line towards the right.

4. In order to find by triangulation a point of the orbit of Mars, take the opposition
of 1920 and draw the positions of the earth and of Mars at this moment (triangle
SE_1M). Look now for the positions of both planets, one Martian year later: Mars has
resumed precisely the same position (look in the column of longitudes, this is the
criterion!). But the earth now occupies another position E_2, which you read in the
table and which you plot.

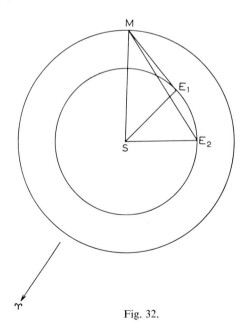

Fig. 32.

5. The angle sun–earth–Mars has been observed (Table I); you are now able to draw the line $E_2 M$ from the earth towards Mars, which intersects $SE_1 M$: the position of Mars in M is thus found.

6. Repeat the same construction at an interval of *two* Martian years after the opposition of 1920. You find again a line $E_3 M$. Actually the three lines should intersect precisely at the same point M; take the best approximation.

7. Instead of starting from the 1920 opposition, take the oppositions of 1924, 1931 (and, if there is time, 1926). For each of them repeat the constructions, described in sections 4, 5, and 6.

8. Draw a circle through the points of the orbit which you have determined: first measure the vector radii, estimate the position of the centre and the length of the radius; then make slight corrections by trial and error. The predecessors of Kepler had made certain *a priori* assumptions about the kinematics of the planet's motion. By the method which he introduced and which we have followed, the orbit has been obtained from observations only and by pure geometrical considerations.

9. From your drawing, measure a, e and obtain a rough value for π, the longitude of the perihelium. Compare with the values in the literature.

(10.) From a derive the period of revolution and compare with your first result (section 2).

(11). Between two consecutive oppositions the planet has moved by $360° +$ an arc $\Delta\omega$. The time intervals correspond to one Martian year + a time interval Δt. Calculate $\Delta\omega/\Delta t$ along the orbit and plot this angular velocity as a function of the heliocentric longitude. Where is $\Delta\omega/\Delta t$ maximum, where minimum?

From a comparison between $\Delta\omega/\Delta t$ and the varying radius vector r, Kepler deduced his second law. Find this relation from your graph yourself.

(12.) Instead of constructing the triangles SEM by drawing, you might have computed their solution by triogonometry. Do this for one or two cases and compare with the result of the drawing.

13. In section 3 we have mentioned that Kepler, before studying the orbit of Mars, first determined the orbit of the earth. Realize that actually, from the positions at one opposition and at intervals of 1, 2, 3,... Martian years you could have compared the radius vector of the earth at several heliocentric longitudes. In principle this was Kepler's method.

References

KEPLER, J.: *Astronomia Nova* (Introduction; chapters 22, 24, 32, 42).
PANNEKOEK, A.: 1961, *A History of Astronomy*, London, chapter: Kepler.

Preparation

For each student: flat sheet of paper, at least 18×18 cm; a good and rather big protractor; drawing instruments; slide rule; table of logarithms.

A25. GEOCENTRIC COORDINATES OF A PLANET

The Problem

To calculate the geocentric coordinates of the planet Mars, for the same day for which we have drawn the relative positions of the sun, the earth and Mars (A21).

The calculation will be made in such a way, that it might be easily extended to a series of other moments ('ephemeris').

From the *Astronomical Ephemeris* we borrow the elements of two orbits for the current year.

	Mars	earth
$\Omega =$	–	–
$i =$	–	–
$\pi =$	–	–
$a =$	–	–
$e =$	–	–

Passage through perihelion: ———

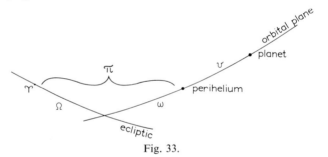

Fig. 33.

Procedure (L)

We apply the formulae, derived in the course of General Astronomy. All calculations are carried out, in 4 decimals, by each of the two partners *independently*; the results are compared after each step, in order to avoid errors. It is absolutely necessary to carry out all operations in an orderly way and to arrange the numbers neatly. Realize at each step what you are doing!

The necessary formulae are summarized on the next page.

1. Determine successively for Mars, at the chosen date:

M (already determined in exercise A21).

E (development in series with respect to M); the terms to be added to M are computed in radians, then reduced to degrees and finally added to M).

v (series with respect to M or to E).

r (from E and a).

2. *Heliocentric equatorial coordinates of Mars.* First compute the constants of Gauss: a, A; b, B; c, C. Each pair of students computes one of the three sets and the results are combined.

Now compute successively:

$$\omega = \pi - \Omega; \quad \text{then} \quad A', B', C'; \quad x_4, y_4, z_4.$$

Compare at this stage your results with the graph, made in exercise A21.

3. *Heliocentric coordinates of the earth.* These would have to be derived by a similar calculation as for Mars. Then their signs would be reversed and they would mean: the geocentric coordinates of the sun. However, since these are directly found in the *Astronomical Ephemeris*, we shall borrow them from there.

Compare again with your drawing of exercise A21.

4. *Geocentric coordinates of Mars.* From X, Y, Z, the geocentric coordinates of the sun, derive the geocentric coordinates ξ, η, ζ $\Big\}$ for the planet.
and $\qquad\qquad\qquad\qquad\qquad \alpha$, δ, Δ $\Big\}$

5. Compare with your earlier drawing. Your present calculation is a considerable improvement, since you have taken into account the elliptical orbit and the inclination of the orbit plane. Compare also with the volume *Planetary Coordinates*; find the distance between your calculated position and the result of the still more precise determination, for which the perturbations were taken into account.

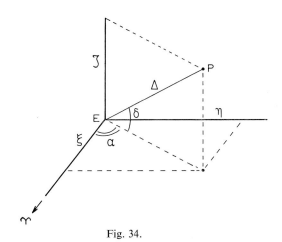

Fig. 34.

Formulae:

$$E = M + e \sin M + \tfrac{1}{2}e^2 \sin 2M + \cdots$$
$$r = a(1 - e \cos E)$$
$$v = M + 2e \sin M + \tfrac{5}{4}e^2 \sin 2M + \cdots$$
or $\qquad\qquad v = E + e \sin E + \tfrac{1}{4}e^2 \sin 2E + \cdots$

For the constants of Gauss:

$$\begin{cases} a \sin A = \cos \Omega \\ a \cos A = - \sin \Omega \cos i \end{cases}$$

$$\begin{cases} b \sin B = \sin \Omega \sin \varepsilon \\ b \cos B = \cos \Omega \cos i \cos \varepsilon - \sin i \sin \varepsilon \end{cases}$$

$$\begin{cases} c \sin C = \sin \Omega \sin \varepsilon \\ c \cos C = \cos \Omega \cos i \sin \varepsilon + \sin i \cos \varepsilon \end{cases}$$

$$\begin{aligned}
A' &= A + \omega & x_4 &= ra \sin(A' + v) \\
B' &= B + \omega & y_4 &= rb \sin(B' + v) \\
C' &= C + \omega & z_4 &= rc \sin(C' + v)
\end{aligned}$$

$$\begin{aligned}
\xi &= X + x_4 & \xi &= \Delta \cos \delta \cos \alpha \\
\eta &= Y + y_4 & \eta &= \Delta \cos \delta \sin \alpha \\
\zeta &= Z + z_4 & \zeta &= \Delta \sin \delta \\
& & \Delta^2 &= \xi^2 + \eta^2 + \zeta^2
\end{aligned}$$

Reference

CLEMENCE, G. M., BROUWER, D., and ECKERT, W. J.: *Co-ordinates of the five outer Planets.* The introduction is quoted in: H. Shapley: 1960, *Source Book in Astronomy 1900–1950*, Cambridge, Mass., p. 93.

Preparation

For each student: table of logarithms (4 decimals); lecture notes, or W. SMART: 1962, *Textbook on Spherical Astronomy* (chapter V); or another book on Spherical Astronomy.

A few copies of the *Astronomical Ephemeris*; *Planetary Coordinates* 1960–1980.

A26. THE THREE-BODY PROBLEM

This problem in its general form is one of the most difficult in celestial mechanics. However, each special case can be solved easily by stepwise numerical integration.

The Problem

We consider a binary star with two components, I and II, which have the same mass and describe circular orbits around their common centre of gravity.

An equally massive star III comes from infinity and passes rather closely along the binary. Which perturbations will these stars exercise on one another? Will one of the components be captured by the passing star? (Origin of the planetary system according to Lyttleton.)

For simplicity we assume that the three stars are moving in the same plane.

BASIC EQUATIONS (Figure 35)

Call m_1, m_2, m_3 the three masses.

Let r_1 be the distance between II and III,
Let r_2 be the distance between I and III,
Let r_3 be the distance between I and II.

According to Newton, the gravitational forces, acting on star I are:

$$G \frac{m_1 m_2}{r_3^2} \quad \text{and} \quad G \frac{m_1 m_3}{r_2^2}.$$

In order to find the resultant forces, we decompose the forces along the X and the Y directions and we make the sum:

$$m_1 \ddot{x}_1 = G \frac{m_1 m_2}{r_3^2} \cdot \frac{x_2 - x_1}{r_3} + G \frac{m_1 m_3}{r_2^2} \cdot \frac{x_3 - x_1}{r_2}$$

$$m_2 \ddot{x}_2 = G \frac{m_2 m_1}{r_3^2} \cdot \frac{x_1 - x_2}{r_3} + G \frac{m_3 m_2}{r_1^2} \cdot \frac{x_3 - x_2}{r_1} \quad \text{etc.}$$

The distances r are always positive. Since the three masses are equal, we may always chose our units in such a way that

$$Gm_1 = Gm_2 = Gm_3 = 1.$$

The equations become:

$$\ddot{x}_1 = \frac{x_2 - x_1}{r_3^3} + \frac{x_3 - x_1}{r_2^3} \qquad \ddot{y}_1 = \frac{y_2 - y_1}{r_3^3} + \frac{y_3 - y_1}{r_2^3}$$

$$\ddot{x}_2 = \frac{x_1 - x_2}{r_3^3} + \frac{x_3 - x_2}{r_1^3} \qquad \ddot{y}_2 = \frac{y_1 - y_2}{r_3^3} + \frac{y_3 - y_2}{r_1^3}$$

$$\ddot{x}_3 = \frac{x_1 - x_3}{r_2^3} + \frac{x_2 - x_3}{r_1^3} \qquad \ddot{y}_3 = \frac{y_1 - y_3}{r_2^3} + \frac{y_2 - y_3}{r_1^3}$$

Procedure (L)

1. Look carefully at these equations and memorize their structure, in order to avoid errors in the computation.

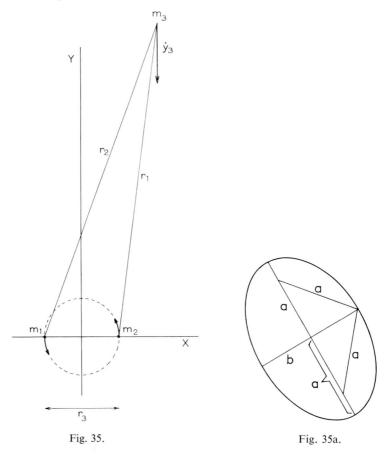

Fig. 35.   Fig. 35a.

2. Choose the following initial conditions:

$$
\begin{array}{llll}
x_1 = -1 & y_1 = 0 & \dot{x}_1 = \cdots & \dot{y}_1 = \cdots \\
x_2 = +1 & y_2 = 0 & \dot{x}_2 = \cdots & \dot{y}_2 = \cdots \\
x_3 = +(2) & y_3 = 10 & \dot{x}_3 = 0 & \dot{y}_3 = (2)
\end{array}
$$

Instead of the dots, substitute those values for which the components of the binary will describe circular orbits. The numbers between brackets may be varied somewhat.

3. Draw the three bodies in their relative positions on scale. Unit of length $=1$ cm.

4. Construct a systematic table, allowing the construction of the orbits, progressing with $t=1$.

5. Measure in the drawing r_1, r_2, r_3, compute the two terms composing $\ddot{x}_1, \ddot{x}_2$, etc. till the 6 accelerations are found for $t=0$. At any moment we should have: $\sum \ddot{x}=0$, $\sum \ddot{y}=0$. This means that the computations may be checked or simplified.

6. Compute the 6 components of the velocities for $t=\frac{1}{2}$, assuming that the accelerations between $t=0$ and $t=\frac{1}{2}$ are still the same as for $t=0$. So

$$(\dot{x}_1)_{t=1/2} = (\dot{x}_1)_{t=0} + \tfrac{1}{2}(\ddot{x}_1)_{t=0}, \quad \text{etc.}$$

7. Compute the 6 coordinates for $t=1$, assuming that the velocity components between $t=0$ and $t=1$ were all the time equal to those, applying to $t=\frac{1}{2}$. So

$$(x_1)_{t=1} = (x_1)_{t=0} + (\dot{x}_1)_{t=1/2}. \quad \text{Etc.}$$

Plot the new positions of the three bodies for $t=1$.

8. From here on, we can improve our method without extra work by introducing a small modification. We assume now:

$$(\dot{x}_1)_{t=1\,1/2} = (\dot{x})_{t=1/2} + (\ddot{x}_1)_{t=1} \quad \text{etc.}$$

Why is this a better approximation than our first step (section 6)? Measure r_1, r_2, r_3 on your drawing. Repeat the operations 5–7, and follow the orbits as long as they are of interest. (In order to save time, the instructor may give you the orbits after $t=6$, say, up to the moment when the disturbing third body has left the stage and when practically the motion has become again a 2-body problem.) Great care should be taken to avoid errors in the computations, especially in the first steps. No more than 2 decimals are taken into account. All computations may be made mentally or with the upper scales of the slide rule.

(9.) Through the successive positions of the three bodies draw fluent lines. Use various colours of ink. Compare with other drawings, constructed for somewhat different initial conditions.

(10.) Of course our calculations are not exact, since we have assumed that the acceleration remains the same between $t=0$ and $t=\frac{1}{2}$. From the second step on we have improved our computation by assuming that $\ddot{x}$ remains the same between $t=\frac{1}{2}$ and $t=1\frac{1}{2}$. By successive approximations this might be improved still further, as explained in books on numerical integration. Even for the specialized astronomer it is surprising how quickly these methods converge.

(11.) If the binary has not been too much perturbed, it is interesting to construct the orbit of the centre of gravity for this system, when the third body has practically disappeared.

(12.) Construct also the orbits of the components with respect to that centre of gravity. These approach ellipses, for which the centre of gravity is the common focus and which are equal, since $m_1 = m_2$; it is sufficient to construct one of them, say that of m_1. This can be done numerically. But you can also calculate the ellipse from the laws of motion as follows. Take one of the points P on the ellipse for which you have already computed the coordinates and the velocity components. Call r_1 the radius vector FP for that point. The corresponding velocity with respect to the centre of gravity is computed from:

$$\dot{\xi}_1 = \tfrac{1}{2}(\dot{x}_1 - \dot{x}_2), \quad \dot{\eta}_1 = \tfrac{1}{2}(\dot{y}_1 - \dot{y}_2),$$
$$v_1^2 = \dot{\xi}_1^2 + \dot{\eta}_1^2$$

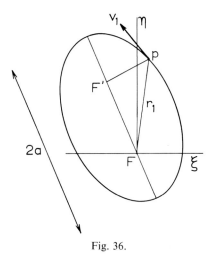

Fig. 36.

The major axis a follows from:

$$v_1^2 = f \frac{m_1 + m_2}{8}\left(\frac{2}{r_1} - \frac{1}{a}\right)$$

(13.) In order to find the second focus F' we first construct the tangent at the same point P of the ellipse, its slope is $\dot{\eta}_1/\dot{\xi}_1$. By a well-known property of the ellipse, the lines $F'P$ and FP form symmetric angles with the tangent, the length PF' is found from the property $FP + F'P = 2a$, (Figure 35a).

In this way the foci, the major axis, consequently also the eccentricity and the minor axis are determined. Sketch the ellipse.

(14.) The period of revolution is proportional to $a^{3/2}$ (Kepler). Compare the period before and after the perturbation.

(15.) Let us calculate the real scale of the system. Assume for the radius of the initial orbits the realistic value 10^{14} cm, which now will be our unit of length. We know that $G = 6.7 \times 10^{-8}$ cm^3 g^{-1} s^{-2}. For $m = 10^{33}$ g the product $Gm = 1$. This will check, if we assume for the unit of time 1.2×10^8 s $= 4$ years, since in those units

$$Gm = 10^{33} \times 6.7 \times 10^{-8} \times \frac{1.4 \times 10^{16}}{10^{42}} = 1.$$

References

BENNETT, A. A., MILNE, W. E. and BATEMAN, H.: 1956, *Numerical Integration of Differential Equations*, Dover Publ.

V. D. KAMP, P.: 1964, *Elements of Astromechanics*, San Francisco.

STERNE, TH. E.: 1960, *An Introduction to Celestial Mechanics*, London.

ZUMKLEY, J.: 1941, *Astron. Nachr.* **272**, 66.

A27. THE ORBIT OF A METEOR

During World War II the city lights had to be extinguished in all Dutch cities and meteor observations could be made better than ever before. Several students, living temporarily at home, decided to make simultaneous observations and succeeded in recording several interesting meteors.

Data

The meteor was observed on August 11, 1944 at 21^h 13 UT. It was observed by 4 observers; their coordinates x, y, z are given in kilometers with respect to Amersfoort, which is taken as origin for geodetical measurements in the Netherlands. The x, y coordinates are measured towards the East and the North in a plane, tangent to the earth's surface in Amersfoort.

Each of the observers drew the observed orbit as accurately as he could in a star map. Later the α, δ coordinates were read from the map and transformed into height h and azimuth A by means of an astrolabe (cf. p. 12).

Observers	Locality	x	y	z
A. Huizing	Hoogeveen	70.9	60.8	−0.7
B. van Woerden	Arnhem	34.3	−18.0	−0.1
C. Businger	Huizen	−12.4	+15.2	0.0
D. Zalmann	Soest	− 5.4	+ 0.8	0.0

	Point of Extinction				Point of Origin			
Observer	A	B	C	D	A	B	C	D
α	$13^h 49^m$	$11^h 24^m$	$11^h 54^m$	$11^h 19^m$	$13^h 14^m$	$10^h 24^m$	$7^h 19^m$	$7^h 39^m$
δ	$31°$	$52°$	$68°$	$66°$	$59°$	$63°$	$79°\frac{1}{2}$	$76°$
A	$102°$	$138°$	$150°$	$150°$	$134°$	$153°$	$178°$	$176°$
h	$31°\frac{1}{2}$	$30°$	$42°$	$38°\frac{1}{2}$	$44°$	$33°$	$42°$	$38°$

Estimated brightness in stellar magnitudes:

$$
\begin{array}{cc}
A & 2^m \\
B & 1^m \\
C & 1^m \\
D & 2^m
\end{array}
$$

The Problem

We have to find the position in space of the two characteristic points: origin and extinction, which determine the orbit near the earth.

Procedure (L)

For simplicity we assume that the four observers were located in the horizontal plane through Amersfoort and that the z coordinates are zero.

    1. Plot the position of the four observers on rectangular coordinate paper (Scale: 1 mm to the km; Amersfoort 3 cm from below).

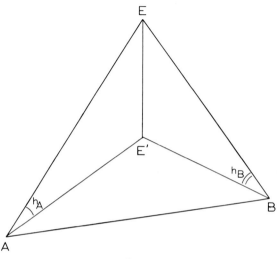

Fig. 37.

    2. We first study the point of extinction, because usually this is the better determined. Plot the four azimuth lines carefully by means of a protractor, or use the value of tan $(A - 90°)$. Take the centre of gravity E' of the intersections; the point of extinction E was located vertically above E'.

    3. Find for each station the observed height $E'E$, making use of the distances AE', BE', ... and the angular heights h. The mean value of the heights determines the position of the point of extinction.

    Small differences between the four heights are inevitable, even with such excellent observations as those reported here. To appreciate their quality, you should try yourself to record a meteor orbit on a dark night!

    4. In general, it is not possible to determine in the same way the point of origin because the meteor flashes up so unexpectedly that often an observer will not have caught the very first appearance. Nevertheless, considering the quality of our data, we shall make an attempt.

    You will find that for three observers the azimuth lines intersect nicely in a point O' and that the height of the point of origin O above O' is well determined. One of the observers apparently noticed the meteor too late, his azimuth line is directed to some point in between O' and E'.

(5.) We shall now check the interpretation, given in section 4. It is possible to draw the orbit of the meteor in space, not trying to find the points of origin and of extinction, but simply the direction of flight.

Imagine a horizontal plane at a height of 100 km, say: the meteor orbit OE intersects this plane somewhere at S. Consider the plane T_1OE, through the observer T_1 and the orbit; it intersects the 100 km plane along O_1E_1. For an observer T_2 a similar line O_2E_2 will exist. Their common intersection is S, which is a point of the orbit.

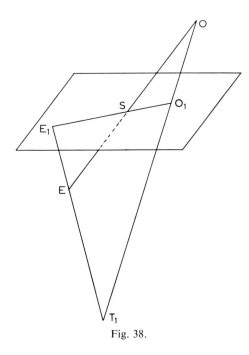

Fig. 38.

For the practical construction we make our drawing on coordinate paper. The azimuth line AE' of the first observer is the projection of the line AE, which, starting from A, ascends by $\tan h$ for each unit of distance along AE'. At a distance 100 km $\times$ $\cot h$ it will have reached a height of 100 km. Plot this point E_1. In the same way, starting again from A and moving towards O', you find O_1. The line E_1O_1 may now be drawn (light pencil line!).

In the same way the observations of B, C, D will yield lines E_2O_2, E_3O_3, E_4O_4. They all should intersect at S; but inevitably there are observational errors and we take the most probable common intersection.

(6.) By repeating the same construction for points at 90 km and 110 km, we find new points $T, U, \ldots$ of the orbit.

(7.) The projections of T, S, U which we have constructed should lie on a straight line, which may be expected to pass also through E, since this had been found with fair approximation. Owing to observational errors some spread of the points is unavoidable and the most probable position of the line has to be estimated.

You may check now whether it passes also through O', which we had assumed to be the projection of the true origin, as determined by three of the four observers.

We have not only the projection of the orbit, we have obtained also a scale of heights along this line. You will be able to tell from what height on the fourth observer has noticed the meteor. Finally draw the orbit to scale in the vertical plane.

8. Compare the heights of origin and extinction with the statistical data in the literature.

9. On the same night, several other meteors were observed. A plot of the observed orbits, as determined by their α, δ coordinates, on a circumpolar map (Norton, map 1–2), shows that the radiant must be located somewhere near Camelopardus or Perseus.

A more precise determination could be obtained by plotting the orbits on a gnomonic map, in which great circles are projected as straight lines.

The date of the observation coincides almost with the maximum frequency of the Perseids. There is no doubt that our meteor belongs to that shower.

References

KATASEV, L. A.: 1957, *Photographic Methods in Meteor Astronomy*, Moscow (Translated from the Russian). Available from the Office of Technical Services, U.S. Department of Commerce, Washington 25 (1964).

MILLMAN, P. M. and McKINLEY, D. W. R.: 1961, in *The Solar System* (ed. by G. P. Kuiper and B. Middlehurst), Chicago, Vol. IV, pp. 677–680 and 721–733.

Preparation

For each pair: rectangular coordinate paper; big protractor; table of trigonometric functions; slide rule; *Norton's Star Atlas*.

Optional: a few gnomonic star maps. The available maps are mentioned by P. M. MILLMAN: 1960, *Trans. Int. Astron. Union* **10**, 351.

A28. PASSAGE OF AN ARTIFICIAL SATELLITE

The impression, made by the passage of the first Soviet Sputnik in September 1957, will remain unforgettable for those who had the chance to observe it. Even now the observation of a man-made satellite of the earth remains an impressing spectacle.

From one of the Satellite Prediction Centres at Cambridge (Mass.), Greenbelt, Moscow or Slough, it is easy to be informed about the coordinating centre for such observations in your country, which will give you the moments of passage of some of the bright satellites. For section 9 a satellite should be selected with an eccentricity <0.10 say.

A telescope or binoculars are only of help for experienced observers and faint satellites.

1L. Let us first test our stopwatches. We stand near the UT clock. At the full minute we start our watch by pressing the button; at the next full minute we stop it, not looking to the hand. The inertia of the observer and of the watch mechanism, if constant, cancels out. Check now whether you have recorded a precise minute. – Repeat this 3 times.

2S. Get a general orientation of the sky. If the orbit has been approximately predicted, notice some of the most important constellations on the track. Prepare your star atlas, damped flashlight, stopwatch.

3S. During the minutes of expectation you will notice how easily one is deceived and thinks that one of the stars is moving. Only by comparison with neighbouring stars can the motion be ascertained. – As soon as one of the observers detects the moving object he signals it, so that all are able to verify the motion.

4S. The problem is now to select a moment when the satellite occupies a well-recognizable position near a star or in between two of them. At that precise moment press the button of your stopwatch, which begins running. Your partner, hearing the click, should also memorize as precisely as possible the position between the stars. – It would be nice if some pairs of observers determined points in the beginning of the passage, others near culmination, and still others near the end.

5L. Directly after the observation note in common agreement this position by a light pencil mark on your star map. Hurry to the UT clock and when it reads a whole minute push again the button of the watch; it stops.

By subtraction you find the UT for the moment when the satellite position was recorded. – For orbit determinations a precision of $0^s.1$ is necessary and is currently reached by experienced observers. By photography $0^s.01$ is attained.

6S. If the satellite is still visible, follow its track. Perhaps you will have the impression of some wobbling: this is an illusion. For some satellites the brightness varies with a period of a few seconds. In other cases you will see it disappear rather suddenly, though it has not reached the horizon: it has entered the earth's shadow.

7L. From your star map read the precise coordinates at the moment which you have recorded (aim at precision 0°.1).

8L. All points observed by the different groups should be plotted on the star map and a smooth pencil line drawn through them. Aberrant points will be directly recognized, corrected or eliminated.

In practice a visual observation with errors of the order 1° in position, 1 sec in time is called 'rough'. With errors of 10′ in position, 0.2 sec in time it is 'semi-precise'.

Very experienced observers are able to record up to 8 moments during one passage and to remember afterwards the 8 corresponding positions!

(9L.) Let us assume that the orbit has only a small eccentricity and may be approximated by the osculating circle. Near the moment of culmination the observer O is looking to the satellite S which moves perpendicularly to the line of sight and to the plane of the drawing in Figure 39; you may determine from your graph section 8 the observed angular velocity $\omega = v/t$. On the other hand you know also the altitude h above the horizon at that moment. You are now able to find the height H at which the satellite moves.

Let the acceleration of gravity be g_0 at the earth's surface, g at the height of the satellite. The equations are:

$$\frac{v^2}{R + H} = g$$

or

$$\begin{cases} \dfrac{\omega^2 t^2}{R + H} = g_0 \left(\dfrac{R}{R + H} \right)^2 \\ (H + R)^2 = R^2 + t^2 - 2tR \cos(90° + h) \end{cases}$$

It is a nice little problem to solve them by successive approximations; H and t are rather small compared with R. Known are: $R = 6400$ km, $g_0 = 0.01$ km s^{-2}, and ω.

We have neglected the velocity of the observer compared to the velocity of the satellite; in how far is this allowed?

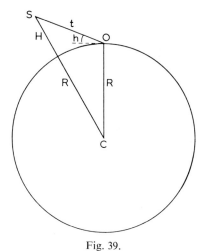

Fig. 39.

Reference

The Optical Tracking of Satellites (ed. by W. de Graaff and C. de Jager), *Cospar Information Bulletin* no. 25, October 1965.

Preparation

For each pair: *Star Atlas*; stopwatch; dimmed flashlight.

PLANETS AND SATELLITES

A29. TOPOGRAPHY OF THE MOON

The Problem

To obtain a general picture of the lunar surface. Especially for the location of the different plains ('Maria'), full-moon is the best moment; this gives also the opportunity to observe some important craters and the systems of crater-rays.

First quarter is the best moment to study some interesting regions in detail. The relief is much more striking, but one observes only part of the surface.

Procedure

1S. A first reconnoitring should be made *with the naked eye*, if possible around full-moon. Sit in an easy position, have a sheet of unlined paper before you, on which you have drawn a circle with a diameter of 10 cm. If necessary, now and then illuminate your paper with a dimmed flashlight (spare the battery!).

Draw with light pencil lines the contours of the dark lunar features, and shade them. Add any other details which you are able to observe. This drawing should not be corrected afterwards!

2S. Use *binoculars*, and observe how surprisingly more details become visible. Avoid shaking, as far as possible, by leaning against a door-post or against a wall.

3S. Now turn to your *experimental telescope*, which has a parallactical mounting and has a fixed position on its pillar. To guide, you have only to move in one coordinate; clamp the telescope very gently, so that it can be moved by light pressure. Always focus; the best focus may be slightly different for the centre and for the outer parts of the field.

Around full-moon, the brightness is so considerable that you will see after-images. It would be useful to insert a shaded glass or a green glass behind the eye-piece.

First enjoy the beauty of the image and get a general orientation. The image is of course reversed, compared to what you observed with the naked eye. Then start drawing, within a circle with a diameter of 15 cm. Orient the sides of your sheet parallel to the cross-wires in the telescope.

Draw the plains, and shade them in conformity with your observations.

Insert the main craters. Near full-moon they appear as bright spots. Note especially the ray-systems. Look especially to Mare Crisium and to Plato and estimate their distances to the limb, remembering that the long axis of Mare Crisium corresponds about to 0.25 of the lunar radius.

Add other details which you are able to observe.

Note on your drawing: orientation, date, hour, type of telescope, magnification.

4L. Now go to the library and compare your two drawings with each other and with a map of the moon (e.g. in *Norton's Star Atlas*). Perhaps you will be disappointed:

it is very difficult to give a correct impression of such unusual forms. Still your sketches are very useful in order to get a general orientation.

Insert in your drawing small numbers in ink, and make a list of names corresponding to these numbers for the plains, craters and other details which you have observed.

In no case should you try to correct the drawing! It is interesting in itself to find out what one is able to observe by the naked eye or with such a small telescope; what kind of error one is liable to make, etc.

Compare the position of Mare Crisium and of the crater Plato with those on the map, and notice the effects of libration in longitude and in latitude. Find in the *Astronomical Ephemeris* the value of the libration coordinates for the day of observation. – Still better: from the *Ephemeris* find the precise moment of First Quarter and observe the position of the crater Ptolemy with respect to the terminator. The *mean* longitude of its centre is 0′5, away from M. Crisium. Its rim is just tangential to the central meridian. By the libration in longitude this position may be considerably altered.

If time is left, choose an interesting part of the lunar surface, go back to the telescope and, with the lunar map in hand, try to distinguish more details.

In between the observations, the moon should be demonstrated to the individual students in succession by means of an *astronomical telescope*.

References

KOPAL, Z.: 1966, *An Introduction to the Study of the Moon*. D. Reidel, Dordrecht, chapter 16.
WILKINS, H. P. and MOORE, P.: 1961, *The Moon*, London.

Preparation

For each pair: experimental telescope; flashlight, damped, card-board with drawing-paper; two chairs; compasses; *Norton's Star Map*.

A few pairs of binoculars; astronomical telescope.

A30. THE SHAPE OF LUNAR CRATERS

The Problem

You will receive a sheet of Kuiper's magnificent *Photographic Lunar Atlas*. This should be handled with great care! We have selected a region not too far from the equator and not too close to the limb, which gives some simplifications in the reduction.

This sheet represents one and the same region under 4 or 5 different directions of illumination. We intend to determine the height of a few crater walls.

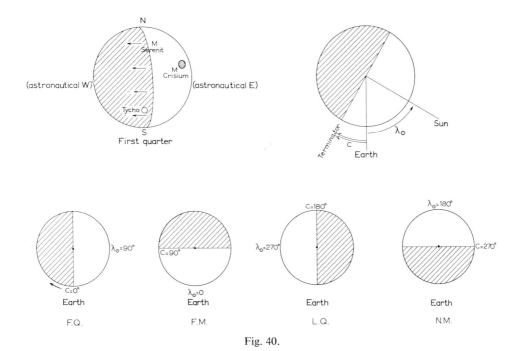

Fig. 40.

Procedure (L)

1. Compare these photographs to the simplified lunar map in your *Star Atlas* or to a survey map. Make some identifications. Enjoy the wealth of details and the beauty of lunar formations.

2. Select a crater which appears suitable for the investigation and find its rectangular coordinates ξ, η, either by measuring them on a survey map, or by looking them up in the *Catalogue* of Blagg and Müller. In this catalogue they are given in units of 0.001 of the lunar radius. The longitude λ of a crater on the equator is given

by the relation: $\xi = \sin\lambda$; more generally the relation is $\xi = \sqrt{1-\eta^2} \cdot \sin\lambda$. Derive λ.

3. The scale of Kuiper's atlas is 2.54 metre to the lunar diameter (3476 km); 1 mm corresponds to 1.37 km. Measure the diameter of the selected crater. In order to avoid perspectivic foreshortening, the diameter perpendicular to the radius of the lunar image should be selected for measurement.

4. We shall now see how the height of a mountain above the surrounding plain may be determined from the length of its shadow. We make a drawing in the plane of the parallel on which the crater is located. Let us provisionally neglect the libration and assume that we are looking from the centre of the earth at an infinite distance.

We suggest that you try to draw a suitable figure and to derive your own formula. Or else you may wish to look at the adjacent diagram and to follow our argument.

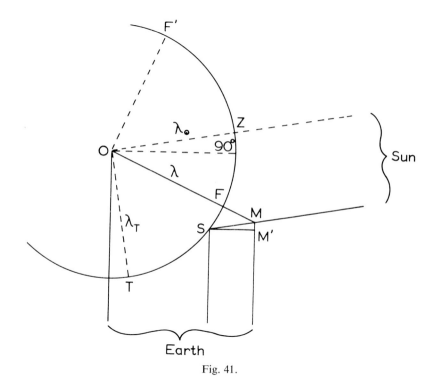

Fig. 41.

We put: O = centre of moon; T = terminator; F = foot of the mountain; M = top of the mountain; Z = subsolar point; S = top of the shadow; the sun has the selenocentric longitude $\lambda_\odot$; the longitude of the mountain is λ; the morning edge of the terminator is $\lambda_T = \lambda_\odot - 90°$; the longitude of other points is indicated along their radii.

The solar rays reach the surface of the moon under an angle

$$MSF = ZOF' = FOT = \lambda + 90° - \lambda_\odot.$$

The shadow thus has a length:

$$SM = \frac{FM}{\sin MSF} = \frac{FM}{\sin(\lambda + 90° - \lambda_\odot)} = \frac{FM}{\sin(\lambda - \lambda_T)}.$$

We see the shadow in projection, it has an apparent length:

$$SM' = SM \cdot \cos(\lambda_\odot - 90°) = SM \cdot \cos \lambda_T = FM \frac{\cos \lambda_T}{\sin(\lambda - \lambda_T)}.$$

Finally the height is found to be:

$$FM = SM' \frac{\sin(\lambda - \lambda_T)}{\cos \lambda_T}.$$

Our result depends strongly on the value of $\lambda - \lambda_T$, which is the distance of the crater to the terminator. The shadows are measured by preference when this distance is small; this means that we should establish the value of $\lambda - \lambda_T$ rather carefully. In practice, lunar observers often introduce the *colongitude* $C = 90° - \lambda_\odot$ (or $450° - \lambda_\odot) = 90° - (\lambda_T + 90°) = -\lambda_T$. Thus the colongitude is the longitude of the morning edge of the terminator, measured from the central meridian *away* from *M. Crisium*. Unlike the selenocentric longitude of the sun, the colongitude *increases* all the time during the lunation.

Consequently:

$$FM = SM' \frac{\sin(\lambda + C)}{\cos C}.$$

5. Take one of the photographs on which the shadows are well visible, neither too short nor too long.

Measure the length of the shadow of the crater rim above the neighbouring crater floor and calculate the height. The longitude λ of the crater has already been found, and C is directly given in the introduction of Kuipers' atlas, table II.

It may be useful to make a simple schematical drawing, indicating the position of the crater, the position of the sun and of the terminator (in the style of Figure 40).

6. Repeat this measurement for another direction of illumination and compare the results; which of the two values has the greatest weight?

7. Draw as well as possible on scale a section of the crater.

(8.) The relation between the diameter d and the depth h below the crater ridge has been studied by MacDonald and later by Baldwin. For craters with $d < 100$ km MacDonald gives the formula $h = 0.378 \sqrt{d}$ or $0.378 \sqrt{d} + 0.95$ in mountain areas (d and h in km). Compare your results with this formula.

References

ARTHUR, D. W. G.: 1962, in *The Solar System* (ed. by G. P. Kuiper and B. Middlehurst), Chicago, Vol. IV, pp. 84–89.

BALDWIN, R. B.: 1949, *The Face of the Moon*, Chicago, chapter 6.

BLAGG, M. A. and MÜLLER, K.: 1935, *Named Lunar Formations*, London.

KOPAL, Z.: 1966, *An Introduction to the Study of the Moon*, D. Reidel, Dordrecht.
   In chapter 14 will be found a much more complete and more precise determination of mountain heights.

Preparation

For each pair: a sheet of Kuiper's *Photographic Lunar Atlas* (Chicago, 1960), for which the colongitude and the selenocentric longitude of the earth are given. *Norton's Star Atlas*, in which a simple map of the moon is found; or other survey maps; trigonometric tables.

N

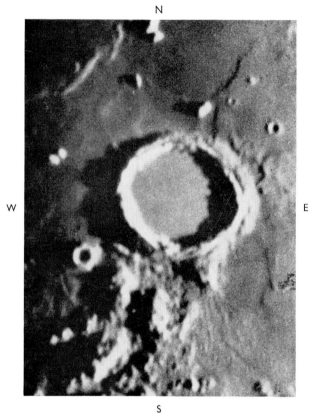

W   E

S

The crater Archimedes, photographed at Pic-du-Midi on 25 october 1963, 19^h03^m UT (Manchester Lunar Programme). $C=8.°2; \beta=+30°; \lambda=-2.°3$ (centre E-wall) and $-5.°3$ (centre W-wall); diameter 80 km. Z. KOPAL: 1966, *An Introduction to the Study of the Moon*, D. Reidel, Dordrecht, p. 202. By courtesy of the author.

The Principle

The stronger the light which reaches a photographic plate, the darker the developed image will be. From the *transparency* u/u_0 of the plate it is thus in principle possible to derive the *radiance i* of the lunar surface, which varies from one point to the other; provided the plate is *calibrated*. We shall have to measure the transparency with a microphotometer (p. XVIII) and to convert this into units of radiance.

Procedure (L)

1. Your simple photoelectric microphotometer should be connected to the A.C. 12 V contact box. The lamp is lighted and its glowing filament is imaged onto a small hole in the stage. The light then reaches a photovoltaic cell and the micro-amperemeter shows a deflection.

2. You have received a reproduction of a lunar photograph, taken with the Yerkes refractor. This plate also carries calibration markings, obtained with a series of known intensities *i*. The exposure-time and the development were the same for the lunar photograph and for the markings, thus they may be directly compared.

3. Put the plate on the stage, gelatine downwards, and first choose a blank part of this plate. You will notice that even this part of the plate reduces already considerably the amount of light transmitted: there is always some *fog*. The condensing lens, which illuminates the plate, can be slightly shifted: note how already small shifts have a considerable influence on the amount of light and thus on the deflection of the micro-amperemeter. Try to get the maximum deflection and then do not change the adjustment any more. (Regulate the sensitivity of the micro-amperemeter so that the deflection corresponds to slightly less than full-scale.)

By inserting for a moment a piece of milkglass near the condensor, you are able to illuminate a larger field and to ascertain where you are measuring.

4. Select a reference part of the blank plate, where you measure the transmission u_0. Then measure the transmission of the density markings u and plot u/u_0 as a function of log *i*. Draw the *transmission curve*.

5. Investigate now the lunar photograph. By comparing with a map you easily identify the main features. Measure successively the transmission at the following points:
   (a) three points inside some of the big plains;
   (b) three points on the brightest mountain regions;
   (c) the bottom of the crater Plato.
For all these measurements you derive u/u_0 (slide rule) and read log *i*.

According to the best measurements the albedo of Plato is 0.068. With respect to this, find the albedo of the other surface elements where you have measured. Between which limits does the albedo vary? Compare with data in the literature.

6. Suppose the photocurrent were not entirely proportional to the illumination u, would this influence your results? And what if the tension on your lamp changed slightly during your measurements?

7. Measure a number of points at random along a radius towards the South Pole, not intersecting big plains; or along a radius away from M. Crisium, and running through plains only. For each point estimate roughly the distance to the centre R/R_0. Plot the radiance of these points. Is there a systematic change of the mean radiance from centre to limb? – How would this be for a smooth white sphere, diffusing according to Lambert's law? Draw the curve.

(8.) It has been found that the brightness of a lunar area is given by: $i = i_0 f(\lambda)$, where i_0 is the albedo at full moon and λ the selenocentric longitude. This means that: (1) the photometric function $f(\lambda)$ is the same all along a meridian; (2) this function is the same for the mountains and for the plains. These rules are very fundamental in lunar photometry.

In order to check them, compare your photograph of the full moon with a photograph of the Last Quarter. Select two points of very different albedo on the same meridian. *Peninsula Fulminum* between *M. Humorum* and *Oc. Procellarum* is a suitable object for comparison with the neighbouring plains on the same meridian. If $f(\lambda)$ is the same for both objects, we may expect: $i/i' = i_0/i'_0$. Check this. – For First Quarter, select the highlands between *M. Crisium* and *M. Foecunditatis*, and compare with these neighbouring plains.

References

MINNAERT, M.: 1961, 'Photometry of the Moon', in *The Solar System* (ed. by G. P. Kuiper and B. Middlehurst), Chicago, Vol. IV, pp. 213–248.
SYTINSKAJA, N. N.: 1953, *Astron. Zh.* **30**, 297–298 (last column).

Preparation

For each pair: microphotometer, connected with 12 V current; milkglass; photograph of the moon, with density markings; map of the moon (e.g. in *Norton's Star Atlas*).

A32. OBSERVATION OF PLANETS THROUGH A TELESCOPE

Especially interesting are the planets Venus, Mars, Jupiter, and Saturn. They are best visible when they are in a favourable phase (Venus) or near the earth (Mars); not too low above the horizon; nor too high, which would make the observation awkward. An astronomical calendar will tell which planets may be observed just now.

OBSERVATIONS WITH OUR LITTLE EXPERIMENTAL REFRACTOR (S)

Always make a simple drawing of what you are observing. Draw also a few background stars, if there are any. Indicate the right orientation: the direction towards which the image is drifting because of the earth's rotation is always West. Note also the date, the hour, the type of telescope and the magnification.

1. Venus: the shape of the phase becomes just visible with our telescope when the apparent diameter reaches 5″; at favourable periods it will exceed 15″.

It may be useful to damp the very bright image by inserting a deep-green glass behind the eye-piece.

2. Mars: details will not be visible; note the reddish colour.

3. Jupiter: most interesting are the four 'Galilean satellites', of which one may be hidden behind the planet. Estimate their position in terms of the diameter of Jupiter, as Galileo did; or by comparing their respective distances. Identify them by means of the *Astronomical Ephemeris*. Observe again the next days and note the motion. Compare their brightness to that of stars and estimate it in stellar magnitudes.

The planet itself shows clearly a disc, which is slightly flattened; the equator coincides with the plane of the satellites. For this observation the deep-green glass will prove useful.

4. Saturn: the ring system will not be visible, it requires at least a magnification of 50.

5. Uranus or one of the brighter planetoids may be observed if they reach the 7th magnitude. Use the *Ephemeris* and the *Star Atlas*, select moments when these objects are easy to find and to identify with respect to bright stars.

OBSERVATIONS WITH AN ASTRONOMICAL TELESCOPE (S)

First note how the telescope is mounted and how it moves. Is there a finder? How is the focus to be adjusted?

The instructor will have to help you, in order to bring the planet in the field. It will spare time if all students consecutively look at the planet for which the telescope has been adjusted.

If there is time available, look first with small magnification, afterwards with a stronger eye-piece. Make a drawing, on which all observable details are reproduced; add all data, as mentioned before.

References

GALILEO: 1610, *Siderius Nuncius*. – Translated into English by E. S. Carlos, 1880; into Italian by M. Timpanaro Cardini, 1947 (with original text). Quotations are found in H. Shapley and H. E. Howarth: 1929, *A Source Book in Astronomy*, New York.

ROTH, G. D.: 1966, *Taschenbuch für den Planetenbeobachter*, Mannheim.

A33. THE ALBEDO OF VENUS

The Problem

The *spherical albedo* of a planet, according to Bond, is the ratio between the total amount of diffused light and the amount of incident light. In order to determine this albedo, we should observe the brightness of Venus from different directions and find the total emitted light.

The Material

We shall make use of the fine measurements of G. Müller, who has measured the brightness of Venus with a photometer according to Zöllner during the years 1900–1909. These measurements have been reduced by his son to 'normal points' and published in the *Astron. Nachrichten* **227**, (1926), 65. They are expressed in star-'magnitudes'.

TABLE

Phase angle α	Brightness m in magnitudes	m reduced to $\Delta = 1$	$\delta m = m - m_0$	$\log I$	f
0°	$- 3^{\mathrm{m}}.75$	$- 4^{\mathrm{m}}.88$	0.00	0.00	1.00
20	$- 3\ .41$	$- 4\ .58$	–	–	–
40	$- 3\ .28$	$- 4\ .24$	–	–	–
60	$- 3\ .59$	$- 3\ .85$	–	–	–
80	$- 3\ .83$	$- 3\ .41$	–	–	–
100	$- 4\ .03$	$- 2\ .94$	–	–	–
120	$- 4\ .23$	$- 2\ .42$	–	–	–
140	$- 4\ .23$	$- 1\ .83$	–	–	–
160	$- 3\ .72$	$- 1\ .14$	–	–	–

Procedure (L)

1. First all measured values of m have to be reduced to what they would be, if the earth was at a distance of $\Delta = 1$ AU from Venus. This has already been done in Müller's publication.

Check this. For the distance Sun–Venus take $R = 0.72$ AU. The distance Δ (Venus–earth) has been found from the heliocentric positions of the two planets $1 = R^2 + \Delta^2 - 2R\Delta \cos\alpha$; Δ will for example be equal to 1 if $\cos\alpha = R/2 = 0.36$ and $\alpha = 69°$. Cf. the table and Figure 42.

2. Now express all brightness numbers m into the brightness m_0 at superior conjunction:

$$m = m_0 + \delta m(\alpha).$$

3. Reduce the magnitudes m to intensities I, considering that

$$\log I = -0.4m_0 + \text{const}.$$

therefore

$$\log I = \log I_0 - 0.4\delta m(\alpha)$$

$$I(\alpha) = I_0 \cdot f(\alpha).$$

4. We shall now consider I and I_0 as the luminous intensity (flux per unit solid angle) emitted by Venus. Similarly let S be the luminance of the Sun; a planet with radius r at a distance R intercepts πr^2. S/R^2. This planet emits:

$$I_0 \int_0^\pi f(\alpha) \cdot 2\pi \sin\alpha \, d\alpha.$$

Consequently the spherical albedo is found:

$$A = \frac{R^2 I_0}{r^2 S} \times \int_0^\pi 2f(\alpha) \cdot \sin\alpha \, d\alpha = p \times q.$$

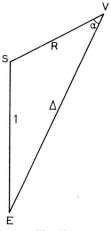

Fig. 42.

5. First determine by graphical integration the integral q. Then compute the factor p, considering that

$$\log \frac{I_0}{S} = 0.4(m_s - m_0),$$

if we compare the brightness of the sun with that of Venus at the same distance of 1 AU.

This gives the albedo of Venus.

6. Study a table with the values of p and q for the successive planets (Russell, H.N. 1916, *Astrophys. J.* **43**, 190.

Numerical Data

Distance sun–earth $= 149.7 \times 10^6$ km $= 1$ AU
Distance sun–Venus $= 108.3 \times 10^6$ km $= 0.72$ AU
Diameter earth $= 12700$ km
Diameter Venus $= 12400$ km
Visual brightness of the sun $m_s = -26^m.72$

Reference

DANJON, A.: 1949, *Bull. Astron. Paris* **14**, 315.

Preparation

For each student: rectangular coordinate paper; table of logarithms.

A34. THE ROTATION OF SATURN

The Problem

The rotation period of Saturn can be determined only on rare occasions from the observation of surface details. A generally applied method, however, is the measurement of the Doppler effect, from which the rotation period is easily derived. The very fine spectrum on which you will work has been obtained at the Lick Observatory on August 19, 1964. The planet was almost in opposition, the ring was inclined by 9°.4 with respect to the line of sight. The spectograph slit was directed along the major axis of the ring (Figure 43).

Procedure (L)

At the upper and at the lower side of the frame the spectrum shows a few bright neon lines, to be used as a comparison spectrum. On the planet itself, the dark lines are clearly slanting, except for a few telluric lines, at the left end, which originate in the terrestrial atmosphere and are perfectly vertical. On the ring you can perceive that the lines are slightly slanting in the opposite direction.

    1. First find the dispersion of the spectrum. Since it was obtained with a grating

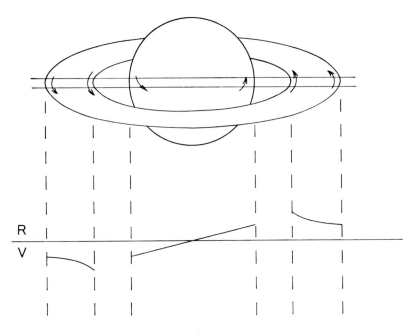

Fig. 43.

it is almost perfectly 'normal'; take lines of the comparison spectrum which are sufficiently far from each other and compute the dispersion in Å/mm.

2. Let one of the partners hold the straight edge of a sheet of paper precisely along a sharp line of the comparison spectrum – top and bottom. The other partner then measures the distance of one of the sharp, dark lines up to that edge:

(a) at the outer limit of the ring ⎱
(b) at the inner limit of the ring ⎰ upper side
(c) at the extreme edge of the planet ⎰

(d), (e), (f) at the same points on the lower side.

Try to reach a precision of 0.1 mm.

These measurements are repeated for at least three other dark lines. Each time the reference edge is again adjusted, in order to minimize accidental errors.

3. In order to find the equatorial rotation velocity v of the planet, we take the mean value of $(c-d)$, first in millimetres, then in Å. By comparing the two opposite limbs, we have doubled the Doppler effect of the motion Sun-Saturn. It is again doubled, because of the effect of the relative motion between Saturn and the earth. Consequently

$$\frac{\lambda_c - \lambda_d}{\lambda} = \frac{4v}{c}.$$

Saturn shows a conspicuous limb darkening. Take care and measure the Doppler shift at the *extreme* limb.

Find the rotation period $T = 2\pi r/v$, where $r = 60400$ km.

Of course this will not completely agree with the number which you find in the literature, since an error of at least 10% in our measurement of $(c-d)$ may be expected.

4. The same measurement is made for the outer and for the inner border of the ring, making use of the differences $(a-f)$ and $(b-c)$. It is clear that the ring does not rotate as a solid body. Each particle moves independently, and $v = \sqrt{fM/r}$. Calculate the mass of Saturn, as well from the outer as from the inner border; both should agree, but for uncertainties in the measurement. Take $f = 6.67 \times 10^{-8}$; $r_A = 139200$ km for the outer boundary of ring A; $r_B = 89600$ km for the inner boundary of ring B.

5. Explain why the lines are straight on the planet. Should they be straight also on the ring?

6. From the Rowland Tables of the solar spectrum you may find the wavelengths of some of the lines in the sun: e.g. Ca 6162.18, Fe 6265.14. Compare with the wavelengths, measured with respect to the neon lines.

At the moment when the spectrum was obtained, the earth was approaching Saturn with a speed of 2.85 km/sec.

References

ALEXANDER, A. F .O'D.: 1962, *The Planet Saturn*, London.
BLANCO, V. M. and McCUSKEY, S. W.: 1961, *Basic Physics of the Solar System*, Reading, Mass.

GINGERICH, O.: 1964, *Sky and Telescope* **28**, 278. Describes the same exercise.

GIVER, L. P.: 1963, *Astrophys. J.* **139**, 727.

SHARONOV, V. V.: 1964, *The Nature of the Planets*, Jerusalem, p. 68 (From the Russian).

Preparation

For each pair: spectrum of Saturn and its rings, to be purchased from Sky Publishing Corporation, 49-50-51, Bay State Road, Cambridge, Mass.

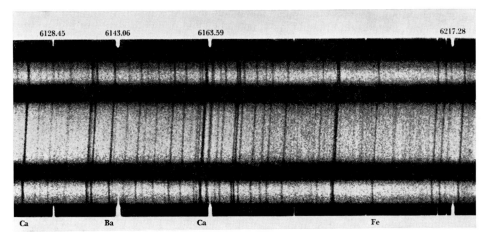

Spectrum of Saturn and its rings, taken at Lick Observatory by H. SPINRAD and L. P. GIVER (1964): *Sky and Telescope* **28**, 278. (Courtesy O. GINGERICH and *Sky and Telescope*.)

B. THE STARS

THE SUN

B1. THE SIZE OF THE SUN (S)

The Problem

In this exercise we shall determine the apparent diameter of the solar disc, which we shall record in radians or in minutes of arc. Given the distance sun-earth, it is easy to reduce this angle to linear measure.

These measurements are easiest to make when the sun is fairly low.

1ST METHOD

1S. In the clip of a stand, place a square of cardboard in which a small hole has been bored. The cardboard should be fixed perpendicular to the rays of the sun.

Catch the beam of light behind the perforation onto a white screen, which you move gradually away from the cardboard. You will observe a bright disc of which the rim is not sharp – clearly, because the aperture is not infinitely small. If we measure the diameter d of the bright disc up to the middle of the transition zone, we get the same value which would be found for an infinitely narrow hole (Figure 44). – Measure also the distance a from the cardboard to the screen (by means of a cord).

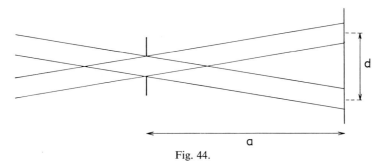

Fig. 44.

2S. Repeat this measurement for another distance between the cardboard and the screen.

3L. Compute from both measurements the solar diameter

$$\gamma \text{ (in radians)} = \frac{\text{diameter bright circle}}{\text{distance cardboard-screen.}}$$

Measurements with such a 'pinhole camera' seem extremely crude. Still, this method has been often used for X-ray photographs of the sun from rockets, since there are neither mirrors nor lenses for X-rays.

2ND METHOD

4S. Now put in the clip another cardboard, carrying in the centre a faintly convex lens. The nominal strength D of the lens in dioptres is written on the cardboard, and gives an approximate measure for the focal distance: $f = 1/D$ meter.

Find as precisely as possible the distance where the solar image is sharp. If this image is too bright, use a grey or a black screen; if the image is sharper at two opposite sides of the solar disc and less sharp in the perpendicular direction, this means that the lens is not quite perpendicular to the rays of the sun.

Put your finger before part of the lens and observe the effect on the image.

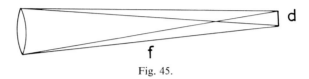

Fig. 45.

Study in detail the image of the sun. Notice the limb darkening and the limb reddening. At the periphery you see little 'waves' running, due to the currents of hot and cold air in the terrestrial atmosphere through which we are looking ('scintillation'). Move the screen quickly to and fro, in its own plane: the grain of the paper is now less troublesome and you might notice sunspots.

5S. Finally measure two diameters, d_1 and d_2, which should be almost equal and of which you take the mean d; measure also the real focal distance f. Compute γ (in radiants) $= d/f$. (Figure 45).

The solar image which you have obtained is certainly not a point. Thus it is wrong to say that 'the solar rays may be considered to be parallel'; the rays, coming *from one point* of the solar disc may be considered to be parallel, but the directions are markedly different for the individual points of the disc.

(6S.) Repeat the measurement with one or two lenses of different focal distances.

(7L.) Compute the mean value $\bar{\gamma} = \sum d / \sum f$; this is a better value than the mean of d/f for now the greater numbers get a greater weight.

(8L.) The weaker the lens, the longer its focal distance and the bigger the solar image. Plot d against f and see whether the plot is a straight line.

Compare your measurements of f with the nominal values $1/D$ and see in how far they agree.

3RD METHOD

Because of the daily rotation of the sky the sun describes a circle around the sky in 24 hours, having an angular extent of $2\pi \cos\delta$, where δ is the declination of the sun (cf. exercise A5). If we determine how much time it takes for the solar image to shift over its own diameter, it will give us a measure of this diameter in radians.

9S. Draw on your screen two perpendicular lines, AB and CD (Figure 46). Adjust it in such a position that the sun moves precisely along AB. Once this direction is found, determine by your watch (in seconds) how long is the interval between the moment when the perpendicular line CD is tangent first to the preceding, and later

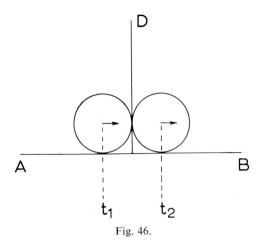

Fig. 46.

to the following limb of the sun. Repeat your measurement at least twice, take the mean Δt of the intervals Δt_1, and Δt_2 and compute $\gamma = 2\pi \cos \delta [\Delta t/(24 \times 3600)]$. The declination of the sun is to be found in the *Astronomical Ephemeris*.

10. Compare the results of the three methods. Which value of γ would you finally adopt?

Reduce γ to minutes of arc (remember: $2\pi = 360°$; etc.).

Compute the diameter of the sun in km, knowing that the distance earth-sun amounts to 150×10^6 km.

Remark

This 3rd method may be applied also by means of our telescope. The solar image is projected onto a screen (cf. B9); it is now bigger, sharp, colourless; the instrument is not shaken by the wind; the crosswires can be used as reference marks.

The method of measurement with a sextant is found in A13.

TABULATION

I.

	d	d ....
	a	a ....
	d/a.........	d/a.........

II.

f_1	d_1	d_2	$\bar{d}$
—	—	—	—
—	—	—	—
$\sum f =$ —	—	—	$\sum d$

$$\gamma = \frac{\sum d}{\sum f} = ...$$

III.

$$\Delta t_1 = \ldots\ldots\ldots \qquad\qquad \delta = \ldots\ldots\ldots$$

$$\Delta t_2 = \ldots\ldots\ldots \qquad\qquad \cos \delta = \ldots\ldots\ldots$$

$$\Delta t = \ldots\ldots\ldots$$

$$\gamma = 2\pi \cos\delta \, \frac{\Delta t}{24 \times 3600} = \ldots\ldots\ldots$$

Adopted value: $\gamma = \ldots\ldots\ldots$ rad. $= \ldots\ldots\ldots$ min of arc.

Diameter of sun $= 150 \times 10^6 \times \gamma$ km $= \ldots\ldots\ldots$

References

KUIPER, G. P. and MIDDLEHURST, B.: 1953, *The Solar System* I, 18.
SCHÖNBERG, E.: 1956, *Sitzungsber. Bayerischen Akad. Wiss.*, p. 243.

Preparation

For each pair: the stand of our altimeter, to which a clamp can be attached (Figure 6); a lower stand for fixing the screen (Figure 7); cardboard square with a small hole; screen, with sheets of white, grey and black paper; several lenses, mounted in the centre of cardboard squares, with focal distances between 1 and 2 m.

B2. THE SOLAR CONSTANT (S)

The Principle

We intend to measure the flux of energy, emitted by the sun, and reaching a brass disc, painted dead black. The radiation is absorbed; a thermometer indicates the rise of temperature of the brass (Figure 47). Our instrument is a simplification of Abbot's 'silver disc pyrheliometer'. The measurement is admittedly rough but it gives the order of magnitude.

The disc of lead or brass is suspended with three strings inside a cardboard tube. This tube is closed at the lower end by translucent paper and mounted with a joint on our stand (Figure 6, p. XXI).

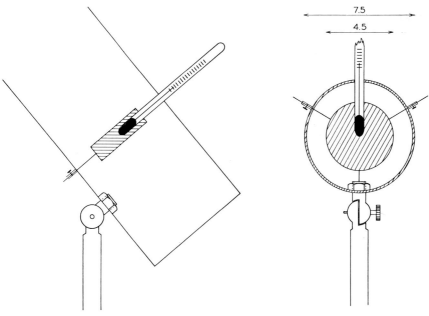

Fig. 47.

Procedure

1S. Select an open terrace and avoid neighbouring walls, heated by the sun. Place the calorimeter in the shade for the present. It is enclosed in a tube of cardboard. Regulate the inclination approximately so that the calorimeter will be directed toward the sun.

2S. Place the thermometer in the hole of the disc, gently pressing and rotating, to ensure a good contact with the metal.

3S. Read the time. Determine the airpath of the solar rays through the atmosphere, expressed in the pathlength of vertically incident rays: put any pencil or rod AB in a vertical position, measure the length of the shadow AC, and compute the ratio $AC/AB = \sec z$ (z = zenith distance).

4S. Now bring the calorimeter into the sunshine. Turn the stand toward the right azimuth; quickly check the inclination, observing the shadow of the disc on the translucid bottom of the calorimeter. Then hold a screen directly before the instrument to protect it from the sunshine.

5S. Read the temperature of the disc in the shadow, at intervals of half a minute, for three minutes. Your partner (A) follows the watch and warns, 5 seconds before the signal 'now!'. At each signal, you (B) read the thermometer, first the tenths of a degree, *then* the whole degrees. (A) records.

6S. Remove the screen at the precise moment of the last signal ('go!') and repeat the temperature measurements again at intervals of half a minute, now, however, in the sunshine.

7S. Again measurements in the shade, as in paragraph 5.

(8S.) Repeat the measurements 3–7 at another solar altitude, later in the afternoon.

9L. When plotting the results, we notice directly that the first half-a-minute does not show the normal temperature increase: this is due to the effects of conduction and other inertia effects. We therefore neglect this first point when computing *the mean rise per minute*:

(a) during the first shadow period (s_0);

(b) during the sunshine period (s);

(c) during the second shadow period (s_0').

The influence of the surroundings, apart from the sunshine, is approximately $(s_0 + s_0')/2$ (either positive or negative). And the corrected temperature increase per minute in the sun amounts to

$$S = s - (s_0 + s_0')/2.$$

10L. Compute the calorimeter constant:

$$G = \frac{\text{specific weight} \times \text{specific heat} \times \text{volume}}{\text{irradiated area}}$$

$$\text{For brass:} \quad \frac{8.5 \times 0.093}{1.2}; \quad \text{for lead:} \quad \frac{11.3 \times 0.031}{1.2}.$$

The smaller this constant, the more sensitive is the calorimeter. The increase of S degrees per minute corresponds to an amount of $K = GS$ calories $\text{min}^{-1} \text{cm}^{-2}$.

(11L.) If the atmosphere did not weaken the solar radiation, K would be the solar constant. However, our measurements, made in the course of the afternoon, clearly show the influence of the airpath. Plot the values of $\log K$ obtained by different groups against the corresponding airpath ($= \sec z$) and draw the nearest straight line (Figure 48). This extrapolated to zero airpath yields the solar constant K_0 and the extinction coefficient A. (Since the extinction is not the same for all wavelengths,

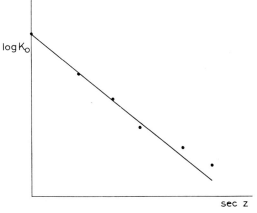

Fig. 48.

extrapolation by a straight line is not entirely justified; correct extrapolation would increase K_0 by about 10%.)

(12L.) How much is the light of a star weakened near the zenith? Express in magnitudes!

(13L.) What is the energy flux emitted by 1 cm^2 of the solar surface? Calculate the effective temperature of the sun from Stefan-Boltzmann's law: $\sigma = 1.37 \times 10^{-12} \, T^4$ cal cm^{-2} sec^{-1}.

TABULATION

Time in min.	t (shadow)	t (sun)	t (shadow)
0	–	–	–
0.5	–	–	–
1	–	–	–
1.5 etc.	–	–	–
	$s_0 = -$	$s = -$	$s'_0 = -$

$\sec z = -$ $S = -$

$G = -$ $K = -$

References

ALLEN, C. W.: 1958, *Commun. Univ. London Obs.* nr. 35.
Annals of the Astrophysical Observatory of the Smithsonian Institution (I–VI).
ROBINSON, N.: 1966, *Solar Radiation*, Amsterdam, p. 94.

Preparation

For each pair: simple pyrhelimeter; measuring rod; cardboard screen.

For each student: rectangular coordinate paper. Table of logarithms.

B3. THE BRIGHTNESS DISTRIBUTION
OVER THE SUN'S DISC (L)

Always handle photographic plates with care! Avoid making scratches! Remember that every finger-print modifies the transmission; thus always take them sideways between your fingers, at the edges.

The Problem

You have received a direct photograph of the sun's disc, either in blue or in red light. You will notice that the density of the plate diminishes towards the limb, corresponding to a limb darkening. This we shall measure quantitatively.

On the same plate there is also a calibration scale, obtained with a tube sensitometer. The illumination for the successive fields corresponds to : 100, 50, 25, 12.5, 6 in relative units.

Record the number of your plate.

Procedure

1. Put the plate cautiously on the plate carrier of the microphotometer, gelatine downward; put a clear part of the plate above the little hole. Switch on your microphotometer, adjust the condenser lens till the micro-amperemeter shows a deflection of nearly 90, then wait a few minutes till the deflection has become constant. Record the deflexion u_0 corresponding to the transmission of the clear plate next to the calibration marks, and measure the transmission $u_1, u_2,...$ in each of these marks. Plot the relative transmissions u_1/u_0, etc. against $\log i$ on logarithmic coordinate paper.

2. Now shift the plate and measure the transmission on points of the solar image along a diameter at distances $r/R = -0.9, -0.8, -0.6, -0.3, 0.0, +0.3, +0.6, +0.8, +0.9$ from the centre.

3. Compute for each point the relative transmission u/u_0 and read the corresponding intensity i from the calibration curve.

Finally refer these intensities to the intensity at the centre of the solar image and take the mean of each pair ($-r/R$ and $+r/R$).

4. Make a plot of i against r/R. This plot suggests that the function might be represented by a formula of the form:

$$i = a + b \cos \vartheta,$$

where ϑ is a heliocentric angle, defined by $r/R = \sin \vartheta$ (Figure 49). In order to test this, plot i against $\cos \vartheta$, see whether a straight line gives a reasonable approximation; find a and b. The ratio $a/(a+b) = (i \text{ (limb)})/(i \text{ (centre)})$ characterizes your curve.

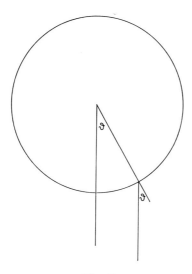

Fig. 49.

(5.) Repeat the procedure for solar images, made in another colour. Or else: compare the amount of limb darkening found by yourself and by other students who worked on different plates.

6. This limb darkening is a consequence of the temperature increase in deeper solar layers. To a limb darkening

$$i_v = a_v + b_v \cos \vartheta$$

corresponds a source function

$$s_v = a_v + b\tau_v,$$

where τ is the 'optical depth'.

TABULATION

$$u_0 = -$$

r/R	i	u	u/u_0	
–	–	–		
–	–	–		
–	–	–	} calibration
–	–	–		
–	–	–		
-0.9	–	–	–	} solar image
-0.8	–	–	–	

References

ALLER, L. H.: 1953, *The Atmospheres of the Sun and Stars*, New York, chapter V.
UNSÖLD, A.: 1955, *Physik der Sternatmosphären*, Springer, chapter II, § 13.

Preparation

For each pair: microphotometer; a direct photograph of the sun's disc, either on a blue-sensitive plate, or on an orthochromatic plate through a yellow filter; calibration by a tube sensitometer. (Such photographs are difficult to obtain in a good quality; a number should be made and the best ones selected.)

Demonstration: the solar image, projected through a telescope.

B4. THE SOLAR SPECTRUM (L)

Introduction

We show you direct photographs of small parts of the solar spectrum, recorded with a big grating spectrograph. These spectrograms are originals and should be handled carefully. They are *negatives*, thus the Fraunhofer lines are bright on a dark background. The wavelength region will be found written on the envelope, in Ångström ($1 \text{ Å} = 10^{-8}$ cm). 'Direct-intensity records' from such plates have been made at the Utrecht Observatory, by means of a special microphotometer. From these you may read how much light is left inside a Fraunhofer line. Do not put pencil marks on these records!

Problem

We have to construct a precise wavelength scale for our intensity record.

Procedure (L)

1. Compare with each other:
   (a) the intensity record;
   (b) the photographic atlas of Rowland (copy of a section);
   (c) the Revised Rowland Table (RRT) or the Second Revision.
Notice the 'wings' of the strong lines; the 'central intensity' inside the lines; the 'blends', where lines are partly superposed.

   2. Somewhere on the intensity record you will find a vertical 'reference line', which may be used as the origin for wavelength measurements (Figure 50). Put on this record a millimeter scale, the reference line coinciding e.g. with mark 50, and the zero point being sufficiently to the left. Make a list of 4 or 5 isolated and well-shaped lines, of which you read, as well as possible, the position x (precision: 0.1 mm). Note also the approximate wavelength from Rowland's atlas; then the exact wavelength y

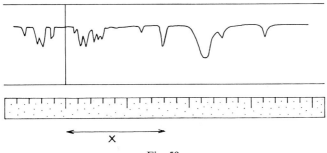

Fig. 50.

on the international scale, from the RRT. Between both columns there is a systematic difference, because Rowland did not know the precise absolute values of the wavelengths.

3. We shall now establish a correct wavelength scale for the whole record. Let x be the measured distances, y the wavelengths; in principle what we have to do is: to plot y as a function of x, then to draw the best-fitting straight line through the measured points. Instead of a graph, we use a simplified 'method of least squares'.

Compute first $\bar{x}, \bar{y}$. This in any case must be a point through which our straight line should pass. The slope of the line is found from

$$D = \frac{\Sigma(x - \bar{x})}{\Sigma(y - \bar{y})}.$$

This is at the same time the spectral dispersion on our record, measured in mm to the Å.

On a strip of strong paper draw very carefully the units of the Ångström scale. The relation between $\bar{x}$ and $\bar{y}$ yields the position of the scale along the record. (This scale should be kept for the next exercise.)

4. Read on your scale the position of some unknown lines and compare with the values in the tables. The difference should not exceed 0.05 Å.

(5.) Look for two lines which are of more or less the same strength and which are just separated on your record. Find in the RRT their wavelength difference $\Delta\lambda$. For an infinitely narrow slit, a monochromatic line and a perfect grating, the *resolving power* $\lambda/\Delta\lambda$ should be equal to mq, where m is the spectrum order, q the total number of lines of the grating. For our spectra $m = 2$; the grating had 600 lines to the millimetre and was 125 mm long. Compute the theoretical resolving power and compare to that really observed.

TABULATION

x	$x - \bar{x}$	λ Rowl.	λ intern. $= y$	$y - \bar{y}$
1	....	....	....	....
2	....	....	....	....
3	....	....	....	....
4	....	....	....	....

$$\Sigma x = \ldots\ldots\ldots \qquad \Sigma y = \ldots\ldots\ldots$$
$$\bar{x} = \ldots\ldots\ldots \qquad \bar{y} = \ldots\ldots\ldots$$
$$D = \frac{\Sigma(x - \bar{x})}{\Sigma(y - \bar{y})} = \ldots\ldots\ldots$$

Check:

λ (from scale)	λ (RRT)	difference
....	....	....
....	....	....

References

BRÜCKNER, G.: 1960, *Photometrischer Atlas des nahen ultravioletten Spektrums*, 2988 Å–3629 Å, Göttingen.

DELBOUILLE, L. et ROLAND, G.: 1963, *Atlas Photométrique du Spectre Solaire de 7498 Å à 12016 Å*, Liège.

MIGEOTTE, M., NEVEN, L., and SWENSSON, J.: 1956, *The Solar Spectrum from 2.8 to 23.7 Microns*, Liège.

MINNAERT, M. G. J., MULDERS, G. F. W., and HOUTGAST, J.: 1940, *Photometric Atlas of the Solar Spectrum*, Utrecht.

MOHLER, O. S. *et al.*: 1950, *Photometric Atlas of the Infrared Solar Spectrum*, 8465 Å–25242 Å, Lake Angelus.

MOORE, CH. E., MINNAERT, M. G. J., and HOUTGAST, J.: 1966, *The Solar Spectrum* 2935 Å *to* 8770 Å, *Second Revision of Rowland's Preliminary Table*, Washington.

Preparation

For each pair: reproduction of a section of the Rowland atlas; direct intensity record of the same section (from the Utrecht Atlas); Revised Rowland Table or Second Revision; strips of strong paper; measuring scale.

B5. PROFILES OF FRAUNHOFER LINES (L)

The Problem

Fraunhofer lines are not really black. Inside each line the brightness varies according to very characteristic laws, connected with the structure of the outer solar layers. Such a 'profile' should be determined on high resolution spectrograms by means of a microphotometer, which measures the transmission inside very narrow strips. Our simple microphotometer does not satisfy such high requirements but will give interesting results just the same.

Procedure

On the plate which you will study, you find (Figure 51):

(a) a calibration spectrum, recorded with a rather broad slit, before which a step weakener had been adjusted; for the successive steps the illumination was proportional to the numbers: 100, 63, 36, 21, 11, 7.

(b) the spectrogram proper, recorded with a much narrower slit. Two or three steps in this spectrum correspond to somewhat different exposure times.

Calibration of the Plate

1. In the first place we have to ascertain the transmission curve of our plate. We follow the general method of photographic photometry, already used in exercise B3. Put the plate cautiously on the plate carrier of your microphotometer (gelatine downward). Connect this instrument to the A.C. contact box shift the plate till the beam passes through a clear part of the plate, and adjust the condenser lens till the microamperemeter shows a convenient deflection (wait a few minutes). – Now look for a place in the spectrum where Fraunhofer lines are almost absent; measure the transmission of the clear plate next to the calibration spectrum (u_0) and that of the continuum in the successive steps $(u_1, u_2, ...)$, shifting the plate in a direction perpendicular to that of the spectrum. Finally measure again the clear plate (u_0'), now at the other side of the spectrum.

2. Compute the mean value $\bar{u}_0 = \frac{1}{2}(u_0 + u_0')$ and the transmissions $u_1/\bar{u}_0, u_2/\bar{u}_0 ....$ Plot these transmission values against $\log i$ on logarithmic coordinate paper (curve *a*).

3. Repeat these measurements, but now in the core of a strong Fraunhofer line of the calibration spectrum, and plot again the transmission curve (curve *b*).

4. The two curves do not coincide, because the intensity of the illumination in the core of the line is considerably smaller than in the continuum. By shifting the curves with respect to each other in the horizontal direction (the direction of $\log i$), it must be possible to bring them into coincidence (why?). In this way you may de-

termine the transmission curve in a more precise way and extend its range. What part of it is the most reliable?

Photometry of the Profile

5. Let us now turn to the spectrogram proper and select one of the very strong lines in one of the steps. Here also we first determine $\bar{u}_0$; then we measure the deflection point by point, shifting the plate in the direction of the dispersion. At great distances from the line rather big intervals may be taken. Notice how surprisingly far the 'wings' of the line may be followed; go on till the deflection on the continuous background has become practically constant.

When shifting the plate by very small amounts, you will notice many small dips in the deflection curve. They will be found to correspond quite nicely with the fainter Fraunhofer lines, superposed on the profile.

6. Reduce your readings to transmission values $u/\bar{u}_0$ and to intensities i. Divide by the background intensity i_c of the continuum and plot the relative intensities i/i_c as a function of the distance Δx to the core of the line. Ask for the dispersion of the plate in mm/Å and compare your profile to that in the *Utrecht Photometric Atlas*.

(7.) You may repeat the operations 5 and 6 for another step of the spectrum. The results must be the same but for observational errors. One of the steps may give better results for the core of the line, the other for the wings.

Do not switch off your microphotometer yet; this might give a fluctuation in the tension and could disturb the work of other students!

(8.) Consider what are the effects, due to which a Fraunhofer line is not perfectly sharp. Have a look into the literature.

TABULATION

Calibration spectrum: $u_0 = \ldots\ldots\ldots$; $u_0' \ldots\ldots\ldots$; $\bar{u}_0 = \ldots\ldots\ldots$

i	u	$u/\bar{u}_0$	u	$u/\bar{u}_0$
0				
100				
65				
36				
21				
11				
7				
0				
	calibration spectrum *a*		calibration spectrum *b*	

Spectrogram proper: $u_0 = \ldots\ldots\ldots$; $u_0' = \ldots\ldots\ldots$; $\bar{u}_0 = \ldots\ldots\ldots$

Δx (mm)	u	$u/\bar{u}_0$	i	i/i_c
$\ldots\ldots\ldots$	$\ldots\ldots\ldots$	$\ldots\ldots\ldots$	$\ldots\ldots\ldots$	$\ldots\ldots\ldots$

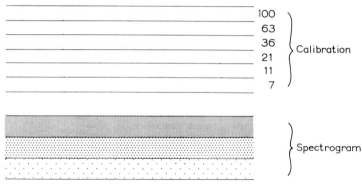

Fig. 51.

References

UNSÖLD, A.: 1955, *Physik der Sternatmosphären*, Kapitel IX, Springer, Berlin.
WRIGHT, K. O.: 1962, in *Stars and Stellar Systems*, Part II, chapter 4, Chicago.

Preparation

For each pair: microphotometer; spectrogram of a small part of the solar spectrum, with a calibration spectrum (as reproduced in the Utrecht *Photometric Atlas of the Solar Spectrum*, Introduction, p. 13); ordinary and logarithmic ($y - \log x$) coordinate paper.

Suitable lines for measurement are $\lambda\lambda$ 5148, 5890, 6563.

B6. PROFILES OF FRAUNHOFER LINES (L)

(Correction and Interpretation)

Problem I

The Fraunhofer line profile which you have determined in the preceding exercise was, of course, somewhat blurred, because the hole of our microphotometer, transmitting the light, is not infinitely narrow. Even with the best instruments there always remains such a blur, due to the imperfect resolving power of the spectrograph: the minima and the maxima are less sharp than they ought to be. The amount of blur depends on the *apparatus function*, which for the Utrecht atlas is found in the Introduction, p. 9–13: it shows how an ideal, monochromatic line is widened by the instrument. (Later it has been shown to be actually narrower.)

Given the *observed profile* in this atlas and the *apparatus function*, we have to disentangle them and to draw the true *profile*.

PROCEDURE

Many methods have been devised for this operation; we shall follow one of the simplest procedures, well suited for the investigation of stellar spectra with many lines. – We consider only the central, most important part of the apparatus function, and as representative we take the centre and the points where the function has decreased to 10% of its central value. It is found that the points, thus defined, are distant by about 0.05 Å towards the red or towards the violet, corresponding with 1 mm to either side. (We have taken round figures.)

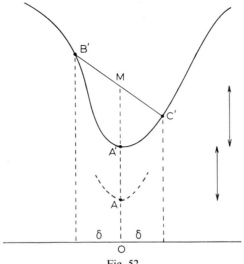

Fig. 52.

Our method of reduction will now be to apply successively to any point $f'(x)$ of the recorded profile the formula:

$$f(x) = 2f'(x) - \tfrac{1}{2}f'(x + \delta) - \tfrac{1}{2}f'(x - \delta),$$

where $\delta = 0.05$ Å; the ordinate $f(x)$ is the function, corrected for the influence of the apparatus. Note that for a horizontal line $f'(x) = f'(x \pm \delta)$ and that the formula then yields: $f(x) = f'(x)$, as it should. The coefficients 2, $-\tfrac{1}{2}$, $-\tfrac{1}{2}$ have been chosen by comparison with other, more theoretical and elaborate methods.

The formula has the following simple geometrical interpretation: $AO = A'O - A'A = A'O - A'M$ (Figure 52).

   If the blur would be simply due to the finite width 2δ of the slit, it is shown that the reduction should amount to: $AO = A'O - (1/3) A'M$. See: Rayleigh, *Scientific Papers*, reprinted 1964 in Dover Publications, New York, Vol. I, 135 (1871).

In our case the wide 'wings' of the instrumental profile have a considerable importance. Our method therefore is more a qualitative illustration than an exact reduction.

   1. Select a few well-shaped lines of different strengths, not showing duplicity. Read and tabulate the ordinates at successive intervals of 1 mm or 2 mm (magnifier!) and draw the profile on a magnified scale.

   2. Now apply to a number of points the simple construction of Figure 52 and draw the corrected profile.

   Pay special attention to the central intensity.

   (3.) Compare your result with that of more refined methods (P. J. Gathier: 1961, *Bull. Astron. Inst. Neth.* **16**, 128).

References

UNSÖLD, A.: 1957, *Physik der Sternatmosphären*, S. 65.
VAN ALBADA, B.: 1937, *Bull. Astron. Inst. Neth.* **8**, 179.

Problem II

The shape of a Fraunhofer line profile may be approximated by the simple formula:

$$\frac{1}{R} = \frac{1}{x} + \frac{1}{R_\infty},$$

where $R =$ dip at a given point of the profile; R_∞ maximum dip for strongest lines in this spectral region (slightly less than 1); $x =$ selective absorption or scattering at this wavelength by the atoms which produce the line. It is easily seen that for small x the term $1/R_\infty$ may be neglected and $x \to R$. While for big x automatically $R \to R_\infty$ as it should; there comes a 'saturation'.

   1. Select a strong, well-shaped line, as e.g. one of the sodium D lines (λ 5890 or

5896) for which the instrumental blur may be neglected. Assume a value of R_∞, taken from this line or from neighbouring stronger lines, if any. Compute $1/R_\infty$.

By subtracting this term, we shall be able, more or less, to eliminate the saturation effect.

2. For a certain number of points on the profile, tabulate in succession $\Delta\lambda$, $1/R$, $1/x$ and x. (Slide rule!).

3. The profile of strong lines is mainly widened by damping; this has been found because x was shown to be proportional to $1/(\Delta\lambda)^2$. Test this by plotting x against $1/(\Delta\lambda)^2$.

References

UNSÖLD, A.: 1957, *Physik der Sternatmosphären*, p. 245 and p. 407.
UNSÖLD, A.: 1967, *Der neue Kosmos*, Berlin, § 19.

Preparation

For each pair: a record from the *Utrecht Photometric Atlas*, including Fraunhofer lines of very different strengths and with a nice, straight, continuous background; the section with the sodium *D* lines is very suitable.

B7. EQUIVALENT WIDTH OF FRAUNHOFER LINES (L)

The Problem

It is necessary to measure not only the wavelength but also the 'strength' of the Fraunhofer lines. This was done by Rowland according to an empirically estimated scale, established by looking at the lines of a spectrogram. The concept of 'equivalent width' gives a more quantitative estimate.

Procedure

1. We shall work again on a section of the Rowland atlas and on the corresponding Utrecht microphotometer tracing, perhaps the same for which you have already constructed the scale in exercise B4. Select a number of well-shaped, isolated lines; look for faint as well as for strong lines. Indicate their position on your scale with crosses, in order to recognize them quickly. Compare with the Rowland atlas and with the RRT, remembering the difference $\lambda_{Rowl} - \lambda_{intern}$. For the identification take into account also the surroundings of the line and the strength. We reject lines which are blended or which are telluric (O_2, H_2O).

(2.) Establish a scale of line-strengths according to your own estimates. The strongest lines are designated by 10, the faintest by 0, and the other ones are interpolated. Try to keep the scale values constant along the whole stretch.

(3.) Compare your scale with that of Rowland: plot the estimated strengths and draw a curve, giving the relation between both.

4. We shall now introduce the concept of *equivalent width* and give a better quantitative definition of the line strength. First put on the intensity record a transparent strip of celluloid, provided with a straight horizontal scratch, which should be made to coincide with the continuous background of the spectrum near the selected line; two little metal blocks will keep the strip in its position (Figure 53).

Now we determine the area of the intensity profile. The record has been printed on a background of square millimetres. We imagine the profile to be subdivided in successive horizontal strips, for each of which we measure the width. Then take the sum and find the total area in mm². We compare this area with that of an entirely dark line, with a sharp rectangular profile; in order to have the same area as the real line and to absorb the same amount of energy, this black line should have a well-defined width which we call the *equivalent width*. The simplest procedure is to compare your line with an imaginary, perfectly dark line, having a width of 1 Ångström. Usually we express the equivalent widths of Fraunhofer lines in *milli-Ångströms*.

Any deformation of the line profile by the spectrograph as studied in B6 leaves the equivalent width unchanged; the energy is simply distributed in a slightly different

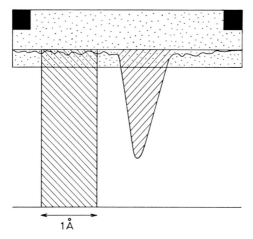

Fig. 53.

way. Consequently, the equivalent width of a line does not depend on the spectrograph.

5. Measure the equivalent width for a number of lines and calibrate the Rowland scale for this part of the spectrum.

How important are the deviations from the mean curve, occurring in Rowland's estimates and expressed in numbers of his scale?

(6.) Compare with the calibration curves, obtained by other students in different parts of the spectrum.

(7.) Areas may be measured mechanically and quickly with a planimeter. Such an instrument should be demonstrated and the principle explained. For faint Fraunhofer lines it is not sufficiently precise.

TABULATION

λ	Estimated strength	Rowland strength	Area in mm^2	Equivalent Width	Difference Rowland-curve

References

MOORE, CH. E., MINNAERT, M. G. J., and HOUTGAST, J.: 1966 'The Solar Spectrum 3935Å to 8770Å, Nat. Bur. St., Monograph 61.

MULDERS, G. F. W.: 1934, Aequivalente Breedten van Fraunhoferlijnen, Nijmegen; 1935, Z. Astrophys. 11, 132.

WEMPE, J.: 1947, Astron. Nachr. 275, 97.

Preparation

For each pair the same as for B4; a celluloid strip with two little metal blocks; magnifier.

Demonstration: the use of a simple planimeter.

B8. THE CURVE OF GROWTH (L)

The Principle

The equivalent width W of a Fraunhofer line is a function of the product Nf. In this product, N is the number of atoms per cm^3 absorbing this particular line, f is the *oscillatory power* of that line, determined from laboratory experiments or from atomic theory.

The *curve of growth* indicates how the equivalent width increases with the number of absorbing atoms.

In order to construct this curve, we would have to look at one and the same line and to modify N in the solar gases which, of course, is impossible. Instead, we might look for lines absorbed by the same atomic level and so having the same N, but for which the factor f is different. Or again we may compare different lines which behave almost as if they were the same, but for which the products Nf are in a well-known ratio: we select components of one and the same spectral *multiplet* of an atom, which are formed in the same wavelength region and having initial levels of about the same energy and final levels of about the same energy. The number of atoms absorbing them are in the same ratio at all heights in the outer solar layers.

Consider e.g. the iron atoms in the solar gases. They are distributed over the several energy states with energies $0, \varepsilon_1, \varepsilon_2, \ldots$. According to a very general law of statistical mechanics (Boltzmann) the number of atoms N in the energy state ε is proportional to $ge^{-\varepsilon/kT}$; g is a small whole number: the *statistical weight*. For two iron lines, absorbed respectively by the levels ε and ε', we have

$$\frac{N}{N'} = \frac{g \exp(-\varepsilon/kT)}{g' \exp(-\varepsilon'/kT)}.$$

If the two iron lines belong to the same multiplet, $N/N' \simeq g/g'$ and $Nf/N'f' \simeq gf/g'f'$.

For such multiplet components the relative values of gf are (in most cases) easily determined theoretically, much better than for arbitrary lines. In Table I they are given for a few multiplets. By comparing several of such lines, for which the products gf are in different ratios, we reach the same effect as would be obtained by changing the number of atoms N. – The multiplet numbers correspond to those in the widely used tables of Mrs. Ch. Moore.

Procedure

1. Take a multiplet of *Ti*. In the table you find W and gf for the component lines. Plot $\log W$ against $\log gf$: this graph is a short segment of the curve of growth (Figure 54).

TABLE IA

gf Values for Solar Multiplet Lines

Ti I

multiplet no.		λ	$\log gf$	W (mÅ)
3		5460.51	−2.36	8.5
		5426.26	−2.48	5.5
		5490.84	−2.67	2.5
4		5210.39	−0.90	86
		5192.97	−0.96	80
		5173.74	−1.06	67
	$v_s = 0.03$	5219.71	−1.90	25
		5152.20	−1.73	38
		5147.48	−1.71	36
5		5064.65	−0.87	79
		5039.96	−0.96	66
		5009.65	−1.96	24
		4997.10	−1.90	27
35		5366.65	−1.89	2.5
		5389.18	−1.69	5
38		4981.73	+0.57	112
		4991.07	+0.45	102
		4999.50	+0.38	104
		5016.16	−0.44	60
		5020.03	−0.29	86
	$v_s = 0.82$	5022.87	−0.30	72
		5024.84	−0.47	62
		5045.40	−1.49	10
		5043.58	−1.30	14
		5040.64	−1.37	16
39		4926.15	−1.71	5.5
		4937.72	−1.72	10
109		5145.47	−0.19	37
		5113.45	−0.36	23
		5087.06	−0.55	22
		5109.43	−0.92	5
	$v_s = 1.44$	5085.33	−1.02	5.5
110		5036.47	+0.30	66
		5038.40	+0.23	60
		5071.48	−0.51	25
		5065.99	−0.65	19
156		5297.24	−0.19	18
157		4885.08	+0.43	53
		4899.91	+0.42	57
		4913.61	+0.33	61
173		5025.57	+0.44	20
		5013.28	+0.31	59
	$v_s = 1.99$	5000.99	+0.30	44
		4989.14	+0.09	29
		4964.71	−0.37	7

Table IA (Continued)

multiplet no.		λ	log gf	W (mÅ)
183		5224.30	+0.42	36
		5224.56	−0.03	68
		5223.62	−0.09	11
		5222.69	−0.05	23
	$v_s = 2.09$	5263.48	−0.27	13
		5247.29	−0.15	10
		5186.33	−0.36	7
		5194.04	−0.08	10
		5201.10	−0.22	11
		5207.85	−0.16	8
200		4921.77	+0.38	40
		4919.87	+0.21	24
		4928.34	+0.27	30
		4948.18	−0.56	7.5
		4941.56	−0.39	3
201	$v_s = 2.15$	4848.49	−0.05	11
		4864.19	−0.37	3
		4880.92	−0.46	9
202		4731.17	−0.10	11
		4733.43	−0.21	11
		4742.13	−0.52	5
216		4995.06	−0.32	3
231		4856.01	+0.66	39
		4870.13	+0.58	36
		4868.26	+0.48	26
		4882.33	−0.16	7
232	$v_s = 2.24$	4778.26	+0.02	16
233		4759.27	+0.72	41
		4758.12	+0.71	40
		4742.79	+0.43	27
		4766.33	−0.14	5
		4747.68	−0.43	4.5
		4734.68	−0.54	3

TABLE IB
gf Values for Solar Multiplet Lines
FeI

multiplet no.		λ	log gf	W (mÅ)
1		5166.29	−3.68	115
		5247.06	−4.50	59
	$v_s = .05$	5254.96	−4.23	92
		5110.41	−3.34	126
		5168.90	−3.49	114
		5225.53	−4.26	68
15		5328.05	−1.43	375
		5405.78	−1.78	266
	$v_s = .97$	5397.13	−1.85	239
		5429.70	−1.76	285
		5446.92	−1.86	238
		5455.61	−2.01	219

Table 1B (Continued)

multiplet no.		λ	log gf	W (mÅ)
16	$v_s = .95$	5051.64	−2.71	111
		5083.34	−2.74	95
		5107.45	−2.78	91
		5123.72	−2.79	101
		4939.69	−3.18	96
		4994.13	−2.90	95
		5041.07	−2.73	112
		5097.74	−2.91	84
		5142.93	−2.72	111
		5151.92	−2.83	100
36	$v_s = 1.56$	5194.94	−1.63	126
		5216.28	−1.65	108
		5107.64	−2.04	97
		5332.90	−2.36	96
		5307.36	−2.46	86
37		5227.19	−0.84	277
66	$v_s = 2.20$	5145.10	−2.40	44
		5131.48	−1.92	72
		5098.70	−1.40	102
		5079.23	−1.45	100
		5250.65	−1.52	104
		5198.71	−1.50	87
114	$v_s = 2.35$	5049.82	−1.00	135
		5273.38	−1.36	104
		4924.78	−1.88	101
		5141.75	−1.57	90
		4848.88	−2.79	33
383	$v_s = 2.99$	5232.95	+0.39	346
		5139.47	−0.05	157
		5192.35	+0.18	176
		5226.87	−0.01	160
		5068.77	−0.59	129
		5139.26	−0.19	137
		5191.46	+0.04	160
687	$v_s = 3.40$	4966.10	−0.30	114
		4946.39	−0.74	113
		4010.03	−0.94	91
		4882.15	−1.10	70
		4963.65	−1.21	48
		4875.90	−1.39	55
		4855.68	−1.33	60
		4843.16	−1.30	67
		4838.52	−1.39	51
		5039.26	−0.89	73
		5002.80	−1.03	85
		4950.11	−1.08	76
		4907.74	−1.33	61
718		5029.62	−1.52	41

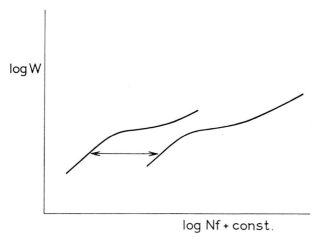

Fig. 54.

2. Repeat this for another multiplet of the same spectral region. The two curves are not coincident: the reason is, that the values of Nf are only *proportional* to gf within each multiplet. If we compare lines of two multiplets, the exponential factors are different. To fit them together we have to multiply all gf values of one multiplet by a suitable factor – this amounts to shifting our curve along the $\log gf$ scale. Copy the first curve on transparent paper, shift it so that the second curve fits onto it and extend your curve. – Add in this way the successive segments from each multiplet, till the whole curve is constructed. Then add the multiplets of Fe, which have stronger lines and define the upper part of the curve.

We may now consider the abscissa as giving $\log Nf$, but for a constant factor. In this way *the curve of growth is the key to a determination of the composition of the solar gases*. Notice how the inclination of the curve is 45° for the faint lines, how it decreases for the medium strong lines, and how it increases again for the strong lines.

(3.) Let us now determine the *excitation temperature* of an element in the solar photosphere. Consider a line, absorbed by the fundamental level, and other lines, absorbed by excited levels of the same element, with considerably different excitation energies $\varepsilon_1, \varepsilon_2 \ldots$ (Figure 54a). For each of these lines we have:

$$\frac{Nf}{N_0 f_0} = \frac{gf}{g_0 f_0} e^{-\varepsilon/kT}, \quad \text{or} \quad \frac{Nf}{gf} = \frac{N_0 f_0}{g_0 f_0} \exp(-\varepsilon/kT);$$

$\log Nf - \log gf = \text{const.} - (5040 \, v_s/T)$ (v_s in electron-volts). (The constant is the same for all lines of that element.) Table I gives the excitation energy v_s for the absorbing levels.

Look up the equivalent widths of a couple of lines in each iron multiplet and read from the curve of growth $\log Nf + \text{const.}$ Find in the table $\log gf$, and plot $\log Nf - \log gf$ against v_s. The slope gives $5040/T$ and thus T.

Why have we compared lines of one and the same atom?

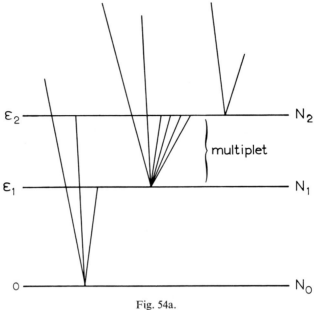

Fig. 54a.

References

ALLER, L. H.: 1953, *The Atmospheres of the Sun and Stars*, New York, ch. 8, Sec. 13.
MINNAERT, M. G. J.: 1953, in *The Solar System*, Vol. I, pp. 150–163.
MOORE, CH.: 1959, *A Multiplet Table of Astrophysical Interest*, Princeton, Washington.
UNSÖLD, A.: 1955, *Physik der Sternatmosphären*, § 106, 107, and Anhang B.
UNSÖLD, A.: 1967, *Der neue Kosmos*, Berlin, § 19.

Preparation

Ordinary graph paper; transparent graph paper.

Data on transition probabilities are found in: Ch. H. Corliss and W. R. Bozman: 1962, *Experimental Transition Probabilities, Nat. Bur. of St. Monograph* **53**, Washington.

B9. SUNSPOTS (S)

The Problem

We intend to observe any spots now visible on the sun, and, if possible, to follow them during several days in order to observe the rotation of the sun. By this rotation all spots are carried along with the same angular velocity (depending somewhat, however, on latitude).

Procedure (S)

1. Put your experimental telescope on its stand. Focus on a very distant object and mark the position of the eye-piece with a pencil mark. Now move the eye-piece 3 mm farther away from the objective: this will enable you to project the solar image onto a screen.

2. Direct the telescope to the sun, but *don't look through it!* Look for the position in which the shadow of the telescope tube has the smallest area. If the direction is right, there appears at once a big bright disc on the screen behind the instrument. Then clamp the telescope gently and follow further the daily motion by moving in hour angle.

3. Draw a circle, 5 cm in diameter on the screen. Put up this screen in a holder behind the telescope at such a distance that the image takes the size of the circle which you have drawn. If the focus is correct, the image of the cross-wires will also appear; the solar image should be centred on them. Focus carefully. Adjust a piece of cardboard to the telescope tube, so that no direct sunlight reaches the screen to confuse the image.

Make sure that the circle, limiting the bright disc, is really the sun's limb and not a diaphragm which limits the field! Remember that we determined the field of your telescope in exercise A5; so there can be no doubt. Moreover, when slightly moving the telescope, the bright solar image must move as a whole by a considerable amount. Perhaps one or two tiny 'spots' will hardly have moved: those are dust-specks on the reticle glass.

4. Now look for sunspots. Rock the paper of the screen quickly to and fro: the grain of the paper seems almost to disappear from sight and we are able to notice even small spots. – Try to draw the spots as precisely as possible, indicating also their size. (The daily motion of the solar image gives some trouble.)

5. In the course of the day the solar image will gradually change its position with respect to the horizon. Thus for comparison, it is necessary to indicate the precise orientation of your drawing as follows:

Do not touch the telescope for a few minutes and mark the direction in which the spots are moving by the daily rotation of the earth. Check that the intersection of the cross-wires remains in coincidence with a pencil mark which you have made on your screen. Now and then you mark on the screen the position of one or two spots. After a couple of minutes you may remove your screen from the stand; draw a straight line through these marks and a parallel line through the centre of the disc, which correspond to the EW direction: W=the direction *towards* which the sun moves; where are now N and S? – Note the date and the hour.

(6.) Project through a bigger telescope some interesting spot groups, observe the umbra and the penumbra.

7. From the observations 4 and 6, derive the real size of the spots in km.

(8.) Repeat the observations 4 and 5 on subsequent days or compare your drawing with that of other students who have observed then. Take account of the orientation and combine all drawings into one. Draw the course followed by some of the spots.

Note. – Drawings of astronomical objects, however simple, are pictures of natural phenomena and must be made with care. They are documents, which may not be changed later!

(9.) If there is a big spot, you may try to observe it by the naked eye, but *only if full precautions are taken!* Use welder's glasses, or a glass plate covered by a thick layer of smoke. Spots bigger than 0.5 or 25000 km were found to be visible for normal eyes. – (Try also when the sun is very low or veiled by fog.) Cf.: *Mem. Brit. Astron. Assoc.* **23** (1921), 19.

References

BRAY, R. J. and LOUGHHEAD, R. E.: 1964, *Sunspots*, London.
SCHEINER, CH.: 1630, *Rosa Ursina*.

Preparation

Note. – Before organizing this exercise, the instructor will have to check whether or not enough spots are visible. If possible, avoid windy weather.

For each pair: experimental telescope; stand with screen (consider the altitude of the sun); cardboard diaphragm, to be adjusted to the telescope.

B10. THE ROTATION OF THE SUN (L)

At Greenwich and at other cooperating observatories the sun's disc is photographed daily and the positions of all spot groups are recorded. This position is determined by the distance to the centre, r, expressed with respect to the solar radius; and by the position angle φ, counted in the sense NESWN, starting from the projection of the polar axis on the sun's disc. (Another method is that of the 'heliographic coordinates', corresponding to latitude and longitude on earth.)

By means of these coordinates we are now able to study the rotation of the sun. Notice, however, that individual spots or groups sometimes may show small deviations from the mean ('proper motions').

Get a general orientation in the Greenwich volume: *Positions and Areas of Sun Spots and Faculae.* The meaning of each column is explained on the first page.

Procedure

1. Draw a circle with a radius of 10 cm, with its horizontal and vertical diameter. The vertical diameter is taken to correspond with the projection of the sun's axis (see the example below).

2. Select a big spot group around December 1 (look at its area). Notice the group number in column 2, and find the same group on the preceding days, starting from the moment when the spot first appeared.

3. Plot carefully the successive positions of this group, by means of a protractor and a measuring scale. Take only every other day, in order to save time.

4. With a smooth line draw the path of the spot group, as it is observed from the earth.

5. Repeat this on a second sheet of paper for another big group which passed the central meridian around March 5 or September 3, and which had a small heliographic latitude. Around these dates the paths have the strongest curvature, showing that then the solar axis is inclined towards us or away from us. Which pole is nearest to us?

6. This inclination i may be approximately determined. Let a and b represent the axes of the elliptical path. Then $b/a = \sin i$. Find this angle from your drawing.

7. Estimate the time interval between the apparition of the spot group at the East limb and the disparition at the West limb. What is the rotation period near the equator?

8. A more precise determination may be obtained from the 'recurrent groups' (Ledger 1, following the *General Catalogue*), as follows:

Note the *number* of one of these groups at each of its successive transits, and look it up in the *General Catalogue.*

Now plot the distance r to the centre of the disc as a function of the date, during the first passage. Repeat this for the following passage (subtracting e.g. 20 days from the date numbers, so that the curves are not too far apart). Measure the horizontal distance between both curves and derive the rotation period.

(9.) Compare your result with that of others who studied spot groups at other latitudes.

(10.) From the 'synodic' rotation period S, you may easily derive the 'sidereal' period T, by taking into account the rotation of the earth around the sun (period E):

$$\frac{1}{T} - \frac{1}{E} = \frac{1}{S}.$$

TABULATION

Spot group No.........

date	r	φ
....	....	....
....	....	....

References

BRAY, R. J. and LOUGHHEAD, R. E.: 1964, *Sunspots*, London.
WALDMEIER, M.: 1955, *Ergebnisse und Probleme der Sonnenforschung*, Leipzig.

Preparation

For each pair: Greenwich Observations, *Heliographic Results* (any volume, but avoid years of sunspot minimum); polar and rectangular graph paper; protractor.

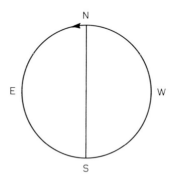

B11. THE SOLAR CYCLE (L)

As an index for the solar 'activity' the *sunspot relative numbers* have been generally adopted: they were defined by Wolf, Wolffer and Brunner, at the Zürich Observatory where they are still now regularly determined on the basis of observational work done over the whole world.

Let G be the number of groups, observed at a certain moment; S the number of spots; then the relative number is $R = 10G + S$. These numbers apply to the original 8-cm telescope with which Wolf started observing in 1852. The results of other instruments are reduced to that same instrument.

Procedure

1. You have received a table, giving the mean relative numbers over a period of 200 years. Plot part of these numbers on a suitable scale.

2. Determine the mean duration of a cycle during this interval.

3. Determine for all available cycles the duration T_m from one minimum to the next. Notice the spread.

4. The magnetic observations suggest that there might be some difference between odd and even cycles. Number your cycles, and determine for each of them:

(a) the maximum value R_m;

(b) the time interval t between the minimum and the next maximum ('time of ascent').

Plot t as a function of R_m, separately for the odd and for the even cycles. Draw smooth curves and discuss them. (Note: Waldmeier prefers to plot $\log R_m$ against t.)

5. Select 3 cycles in which R_m is respectively big, medium and small. Your notes obtained in the preceding paragraph may be helpful for such a selection. Draw these three cycles in such a way that the maximum years coincide. Compare your results with those of the preceding paragraph. *These cycles form a family of curves, with one parameter.*

(6.) It is also interesting to plot the *daily* relative numbers for a period of 2 months. The effect of the 27-day period is noticeable.

TABULATION

time of min.	time of max.	duration T_m	maximum R_m	Time of ascent t

References

Astronomische Mitteilungen, Zürich.
GLEISSBERG, W.: 1952, *Die Häufigkeit der Sonnenflecken*, Berlin.
WALDMEIER, M.: 1955, *Ergebnisse und Probleme der Sonnenforschung*, Leipzig.
WALDMEIER, M.: 1961, *The Sunspot Activity in the Years 1610–1960*, Zürich, pp. 20, 21.

Preparation

For each pair: table of yearly relative numbers over 200 years; table of daily relative numbers over a few months (for a maximum year); or plots of these numbers (Waldmeier, *loc. cit.*); graph paper.

TABLE I

Relative Sunspot Numbers (After WALDMEIER, 1961)

Year	R	Year	R
1744	5	1779	125.9
1745	11	1780	84.8
1746	22	1781	68.1
1747	40	1782	38.5
1748	60	1783	22.8
1749	80.9	1784	10.2
1750	*83.4*	1785	24.1
1751	47.7	1786	82.9
1752	47.8	1787	*132.0*
1753	30.7	1788	130.9
1754	12.2	1789	118.1
1755	9.6	1790	89.9
1756	10.2	1791	66.6
1757	32.4	1792	60.0
1758	47.6	1793	46.9
1759	54.0	1794	41.0
1760	62.9	1795	21.3
1761	*85.9*	1796	16.0
1762	61.2	1797	6.4
1763	45.1	1798	4.1
1764	36.4	1799	6.8
1765	20.9	1800	14.5
1766	11.4	1801	34.0
1767	37.8	1802	45.0
1768	69.8	1803	43.1
1769	*106.1*	1804	*47.5*
1770	100.8	1805	42.2
1771	81.6	1806	28.1
1772	66.5	1807	10.1
1773	34.8	1808	8.1
1774	30.6	1809	2.5
1775	7.0	1810	0.0
1776	19.8	1811	1.4
1777	92.5	1812	5.0
1778	*154.4*	1813	12.2

Table 1 (Continued)

Year	R	Year	R
1814	13.9	1866	16.3
1815	35.4	1867	7.3
1816	*45.8*	1868	37.6
1817	41.1	1869	74.0
1818	30.1	1870	*139.0*
1819	23.9	1871	111.2
1820	15.6	1872	101.6
1821	6.6	1873	66.2
1822	4.0	1874	44.7
1823	1.8	1875	17.0
1824	8.5	1876	11.3
1825	16.6	1877	12.4
1826	36.3	1878	3.4
1827	49.6	1879	6.0
1828	64.2	1880	32.3
1829	67.0	1881	54.3
1830	*70.9*	1882	59.7
1831	47.8	1883	*63.7*
1832	27.5	1884	63.5
1833	8.5	1885	52.2
1834	13.2	1886	25.4
1835	56.9	1887	13.1
1836	121.5	1888	6.8
1837	*138.3*	1889	6.3
1838	103.2	1890	7.1
1839	85.7	1891	35.6
1840	64.6	1892	73.0
1841	36.7	1893	*85.1*
1842	24.2	1894	78.0
1843	10.7	1895	64.0
1844	15.0	1896	41.8
1845	40.1	1897	26.2
1846	61.5	1898	26.7
1847	98.5	1899	12.1
1848	*124.7*	1900	9.5
1849	96.3	1901	2.7
1850	66.6	1902	5.0
1851	64.5	1903	24.4
1852	54.1	1904	42.0
1853	39.0	1905	*63.5*
1854	20.6	1906	53.8
1855	6.7	1907	62.0
1856	4.3	1908	48.5
1857	22.7	1909	43.9
1858	54.8	1910	18.6
1859	93.8	1911	5.7
1860	*95.8*	1912	3.6
1861	77.2	1913	1.4
1862	59.1	1914	9.6
1863	44.0	1915	47.4
1864	47.0	1916	57.1
1865	30.5	1917	*103.9*

Table I (Continued)

Year	R	Year	R
1918	80.6	1943	16.3
1919	63.6	1944	9.6
1920	37.6	1945	33.2
1921	26.1	1946	92.6
1922	14.2	1947	*151.6*
1923	5.8	1948	136.3
1924	16.7	1949	134.7
1925	44.3	1950	83.9
1926	63.9	1951	69.4
1927	69.0	1952	31.5
1928	*77.8*	1953	13.9
1929	64.9	1954	4.4
1930	35.7	1955	38.0
1931	21.2	1956	141.7
1932	11.1	1957	*190.2*
1933	5.7	1958	184.8
1934	8.7	1959	159.0
1935	36.1	1960	112.3
1936	79.7	1961	53.9
1937	*114.4*	1962	35.0
1938	109.6	1963	27.9
1939	88.8	1964	10.2
1940	67.8	1965	15.1
1941	47.5	1966	47.0
1942	30.6		

B12. THE CONVECTION CELLS OF BÉNARD (L)

The solar granulation is due to *convection cells* in the upper photosphere. As far back as 1901 Bénard showed in the laboratory that a horizontal liquid sheet, heated from below, tends to separate into a great number of 'cells'; in the centre of each of them hot liquid ascends, it cools near the surface, and along the periphery the cooler and heavier liquid descends. In order to see the boundaries between the cells, Bénard used a refined 'schlieren-method', revealing minute differences in refracting index.

We shall repeat this experiment in a simplified form. To enjoy it fully, each student (or each pair) should carry it out himself and look for himself. The experiment is made in the chemistry laboratory.

Procedure

1. We first have to heat and to melt some paraffin wax in a tin cup, *A*. Such a cup is placed upon a strip of aluminium, which in its turn rests on wire-gauze over a tripod.

Check with a level whether the aluminium strip is horizontal. If not, adjust by inserting small pieces of metal or cardboard under the legs of the tripod. On the table have also a smaller cup *B*, which will be used as a support when cup *A* has to cool off. Adjust this cup also till its upper rim is horizontal.

2. If Cup *A* does not already contain solid paraffin from an earlier experiment, cut with your pocket-knife enough chips of paraffin to make a layer, about 4 mm thick (when molten). Heat this cup gently on the aluminium strip with a Bunsen burner till the paraffin begins to melt. Then remove the flame and wait till the paraffin is liquified completely. Cautiously remove the cup from the tripod and put it on cup *B*, taking care to keep it horizontal. (Spare your fingers! If necessary use a small pair of tongs.)

3. Allow the paraffin to cool off until it develops a thin skin of solid paraffin. Wait a few seconds more, till the solid layer is entirely opaque and white. Then transfer the cup to the aluminium strip again, which probably is still sufficiently hot. (It may be necessary to heat it slightly first with the flame.)

Look intently at the paraffin. Within perhaps half a minute or one minute the heat of the aluminium strip has reached the molten bottom layer of the paraffin and the convection cells begin to form. *The solid paraffin melts first above the centre of each cell and by this the cells become visible.*

The surface begins to look patchy; especially when looking vertically down you notice during some seconds a surprisingly regular pattern of cells, which have a size of the order of 5 mm. This pattern is not entirely uniform over the whole surface, due to slight differences in depth and imperfect contact with the aluminium strip. Soon there will be places where the surface layer has wholly melted away; at the boundary of such an area you see the walls of the cells gradually disappearing.

Estimate roughly the distance λ between the centres of two contiguous cells. The ratio between the depth ε of the layer and λ was found by Bénard to be about 0.3, but this applied to another substance.

(4.) When the whole cup is molten again, repeat the operations 2 and 3 with slight variations. For example you may heat the aluminium strip somewhat more before starting operation 3. Or you may wait somewhat longer. Or you may increase the thickness of the layer and notice how the size of the cells increases.

Note. – The cells which we have been watching are a very stable type of convection. In other circumstances they change their shapes and positions all the time; this is the type which resembles the solar granulation best.

References

BÉNARD, H.: 1901, *Ann. Chim. Phys.*, 7e Ser., **23**, 62.
RAYLEIGH: 1916, reprinted in *Scientific Papers*, Dover Publications, New York, 1964, Vol. VI, p. 432.

Preparation

The exercise should be made in the chemistry laboratory. For each pair: tripod with wire-gauze; Bunsen burner; a cup A of thin metal, as used for medicinal tablets, with a flat bottom; diameter between 6 and 10 cm, say. A smaller cup B; a strip of aluminium, 10 cm × 20 cm, thick about 2 mm; paraffin wax; level; any simple measuring rule.

B13. THE SHAPE OF THE CORONA (L)

It is of general knowledge, that the shape of the corona depends on the phase of the solar cycle. The light distribution in the corona is determined in recent years by accurate photometric measurements. However, it is possible to utilize a much greater material, including records of earlier times, by the following simplified method.

Procedure

1. You will receive a set of photographs, obtained at one and the same eclipse. If the reproduction shows rather strong contrasts, it is not too difficult to follow the 'outline' of the corona. This is still easier if you cover the photograph by a piece of tracing paper. Draw carefully such an outline, not paying attention to small irregularities. This is a line of constant brightness, an *isophote*; how the reproduction has been made does not matter.

2. Other photographs of the same corona, for which the exposure time has been shorter or longer, will yield isophotes nearer to the sun's limb or farther away.

Indicate on each drawing the position of the sun's axis.

3. A characteristic parameter for each of these isophotes is the amount of flattening. In order to avoid local irregularities, we apply on our drawing a celluloid sheet, on which some diameters have been engraved according to Figure 55.

Record in millimetres the length of the diameters I, II, III, IV, V, VI and compute the mean equatorial diameter $E=(I+II+III)/3$, the mean polar diameter $P=(IV+V+VI)/3$ and the flattening $\varepsilon=(E-P)/P$.

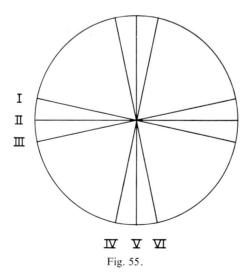

Fig. 55.

Also measure the sun's diameter M and express the equatorial diameter of the corona in solar diameters: $R=E/M$.

4. The flattening changes more or less systemically for increasing distance from the sun's limb. Plot ε as a function of R. See whether you can extrapolate the first part of the plot by a straight line and read ε^* for $R=2$.

(5.) Plot ε^*, obtained by several groups from different eclipses, as a function of the phase in the solar cycle.

6. The way in which ε varies with R is better understood if we remember that the corona, as observed, is the superposition of two phenomena: (1) the light scattered by the free electrons in the highly ionized, rarefied gas around the sun (K corona); (2) the light, diffracted by the cloud of interplanetary dust particles (F corona). By ingenious methods the two components can be separated. The first component varies in shape and intensity with the solar relative number; the second is more or less constant and symmetrically distributed around the sun.

The subsequent table gives the mean intensities which Van de Hulst derived from a great number of eclipse observations, as well along the polar as along the equatorial direction. When the sun is very active, the intensity distribution of the K corona is the same in both directions, and so of course is also the intensity distribution of $K+F$, while $\varepsilon\simeq0$ at all values of R.

Near minimum the situation is different. Derive from the table how the value of ε in this case changes as a function of R, according to the model of Van de Hulst. Compare with the graphs, obtained by you and by other students for different eclipses.

Derive also from the table how ε in minimum periods would vary with R if we could eliminate the interplanetary dust and observe the pure K corona.

It will now become clear, that the decrease of ε at large distances from the sun is due to the F corona, which there becomes predominant.

TABULATION

	Photograph 1	2	3
I	....	....	....
II	....	....	....
III	....	....	....
E	....	....	....
IV	....	....	....
V	....	....	....
VI	....	....	....
P	....	....	....
ε	....	....	....
M	....	....	....
$R=E/M$	....	....	....

References

LUDENDORFF, H.: 1928, *Sitzungsber. Akad. Berlin* **16,** 185.
LUDENDORFF, H.: 1934, *Sitzungsber. Akad. Berlin* **16,** 200.
VAN DE HULST, H. C.: 1953, in *The Solar System* (ed. by G. P. Kuiper and B. Middlehurst), Chicago, Vol. I, p. 285.

Preparation

For each pair: Some photographs of the corona at one and the same eclipse; photographs of very different exposure should be collected from the references, found in the publications of Ludendorff and Van de Hulst (p. 263).

If possible, eclipses near minimum or shortly before should be included (1900, 1914, 1922, 1923, 1943, 1944, 1945) and data should be given up to $R = 2$.

The position of the sun's axis should be marked on each photograph; sheets of tracing paper; celluloid sheet, on which the lines of Figure 55 are engraved.

This exercise will make clear to the students a fundamental principle of photographic photometry: areas of equal brightness remain equally bright after any reproduction process. Methods by which isophotes may be quickly obtained are discussed by P. de Gregorio, *et al.* in: 1967, *Mem. Soc. Astron. Ital.* **38**, 33.

TABLE

Brightness of the Corona (After VAN DE HULST: 1953, in *The Solar System*, I, p. 262.)

R	log K + F	log K
		Maximum, Equator and Pole
1	4.74	+ 4.73
1.03	4.58	+ 4.56
1.06	4.42	+ 4.40
1.1	4.24	+ 4.21
1.2	3.87	+ 3.82
1.4	3.34	+ 3.26
1.6	2.96	+ 2.84
1.8	2.65	+ 2.49
2.0	2.40	+ 2.18
2.5	1.95	+ 1.56
3.0	1.66	+ 1.10
3.5	1.45	+ 0.77
4.0	1.28	+ 0.52
5	1.02	+ 0.17
6	0.81	− 0.06
8	0.50	− 0.40
10	0.26	− 0.64
		Minimum, Equator
1	4.50	+ 4.48
1.03	4.34	+ 4.31
1.06	4.19	+ 4.15
1.1	4.01	+ 3.96
1.2	3.65	+ 3.57
1.4	3.14	+ 3.01
1.6	2.77	+ 2.59
1.8	2.50	+ 2.24
2.0	2.27	+ 1.93
2.5	1.86	+ 1.31
3.0	1.60	+ 0.85
3.5	1.40	+ 0.52
4.0	1.25	+ 0.27
		Minimum, Pole
1	4.35	+ 4.31
1.03	4.15	+ 4.10
1.06	3.97	+ 3.90
1.1	3.75	+ 3.65
1.2	3.29	+ 3.08
1.4	2.73	+ 2.26
1.6	2.42	+ 1.72
1.8	2.21	+ 1.33
2.0	2.04	+ 1.00
2.5	1.74	+ 0.32
3.0	1.52	− 0.24
3.5	1.35	− 0.70
4.0	1.20	− 1.10

On the same scale the sun's brightness, average of disc, corresponds with $\log I = 10.0$.

B14. RADIO BURSTS AFTER A SOLAR FLARE (L)

After a solar flare we observe: (a) several bursts of type III; (b) in some cases a burst of type II. These phenomena become understandable when the radiospectrum is recorded: this is done with a cathode-ray tube on a 35 mm film.

Figure 56 is a section of such a film, reproduced in its true size; the scale of Figure b is 10% smaller than that of Figure a. The frequency scale of the instrument has been determined once for all (Table I); plot the frequency against the deflection for both figures. – The scale is small and the phenomena are complex. – Sorry! These are among the very best records.

Procedure

TYPE III BURSTS (FIGURE 56b)

On this film you notice 3 bursts, of which the left one shows a harmonic duplication. In each of them the frequency quickly diminishes with time. For our study we select the lowest one, which can be followed over the longest distance.

1. Determine as well as possible the coordinates for some points of the streak of light, always looking for the brightest parts and the lowest frequencies. Towards the end some continuum emission comes in and the position of the burst becomes uncertain. Do not pursue your measurements too far!

2. Reduce the coordinates to millimetres and frequencies (Table I).

3. The frequency decrease is explained: a stream of quick particles shoots upward through the corona and excites radio vibrations in the successive layers. A gas with N electrons per cm^3 radiates in the *plasma frequency* $v = 9.0 \times 10^{-3} \sqrt{N}$ MHz. From the measured frequencies you are now able to derive the electron concentration in the successive emission centres. You find numbers which are quickly decreasing.

4. On the other hand the brightness distribution in the corona informs us about the way in which N decreases with height. Take the necessary data from Table II and make a graph. You are now able to find the height H which corresponds to the values of v and N.

5. Finally draw the curve, showing how the height of the 'disturbance' increases as a function of time and determine the velocity of ascension. This disturbance actually is a stream of upward-shooting electrons.

(6.) Note: From direct interferometric measurements Wild (1959) has found that the mean velocity of ascension is near $0.40\ c$ (c = velocity of light). If you have derived a lower value (as most observers did), this shows that in a disturbed region of the corona the density gradient is smaller than in the normal corona. What would your result be if you assume that the disturbance follows a coronal streamer and if you apply the electron densities proposed by Weiss (Table II)?

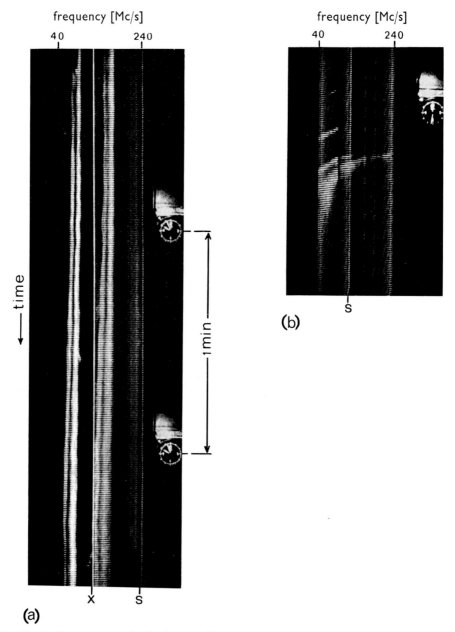

Fig. 56. Radio spectrum of solar bursts. After WILD, J. P., MURRAY, J. D., and ROWE, W. C.: 1954, *Australian J. Phys.* **7**, 456.

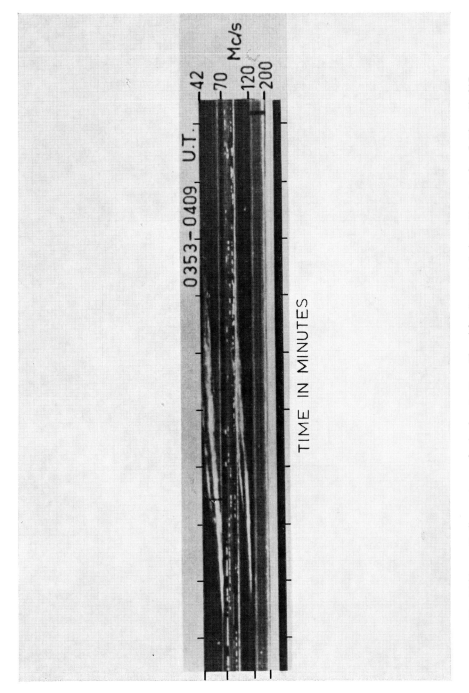

Fig. 56 bis a.   Radio spectrum of solar bursts, Type II. – After ROBERTS, J. A.: 1959, *Australian J. Phys.* **12**, 327.

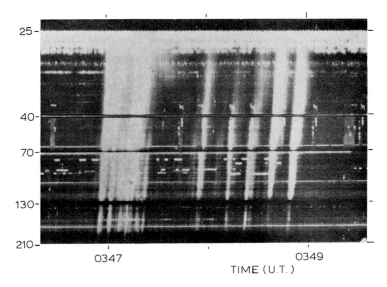

Fig. 56 bis b.   Radio spectrum of solar bursts. Type III (notice quasi-periodicity). – After WILD,
J. P.: 1963, in *The Solar Corona* (ed. by J. W. Evans), p. 115.

TABLE I

| λ (m) | ν (MHz) | deflection (mm) | |
		Fig. a	Fig. b
7.50	40	0.0	0.0
6.00	50	2.3	2.1
4.30	70	7.1	6.4
3.00	100	11.8	10.7
2.31	130	15.4	13.9
2.00	150	17.2	15.5
1.67	180	19.7	17.8
1.43	210	21.6	19.5
1.25	240	23.0	20.7

TABLE II

Electron density in the corona

$\rho = r/r_0$	h (km)	$\log N$	$\log N$(streamer) after Weiss
1.011	7.500	9.05	
1.014	10.000	8.6	
1.03	21.000	8.2	
1.1	70.000	7.9	9.1
1.4	280.000	7.1	8.3
2.0	700.000	6.25	7.4

TYPE II BURTS (FIGURE 56a)

Here the frequency decrease is considerably slower. The film section under investigation is only a small part out of the total record.

7. The burst occurred simultaneously on 4 frequencies. By careful scrutiny you will find that small irregularities and bends are seen on all four of them. The values of the frequencies suggest that one pair might be the first harmonic of the other pair. The great similarity indicates that probably the fundamental frequency, as well as the first harmonic, are split for some reason, perhaps by a magnetic field (Zeeman effect).

For our measurements we select the lowest frequency.

Try for some frequencies whether the ratio 2:1 between the fundamental and the first harmonic is verified.

8. Measure the gradual shift at several moments of time, first in millimetres, then in frequency, in electron density, finally in height.

9. Plot the position of the disturbance as a function of time and compute the velocity of ascension.

For type II bursts it is assumed that the exciting disturbance is a magnetohydrodynamic shock-wave, emitted at the flash-phase of the flare.

(10.) You may similarly study the Figures 56 bis, which are sections of other films, showing bursts of types II and III. In these records the time increases from left to right.

Note. – We have tacitly assumed that the disturbance ascends vertically along a solar radius. It might also follow a skew direction; what would the consequence be as to the velocity, derived from the measurements?

TABULATION

Time (mm)	Time (sec)	Deflection	v (MHz)	$\log v$	$\log N$	h (km)
....	....	....	....	....	....	....

Reference

KUNDU, M. R.: 1965, *Solar Radio Astronomy*, Interscience Publ., New York. See the spectra, notice especially the graphs pp. 297, 324, 325, 342, 357, 374, 376, 378.

THE STARS

B15. TO DRAW A CONSTELLATION (S)

When coming from a well-lighted room, your eyes must gradually adapt themselves to the darkness of the night; this process requires at least a quarter of an hour. As it proceeds, you will gradually enjoy the beauty of the starry sky.

Notice the strong differences in brightness between the stars.

Notice also the colour differences, which are best visible for the brighter stars: white are Sirius, Vega, Spica, Rigel; yellow are Capella and Arcturus; orange-red are Betelgeuse, Aldebaran, Antares, β Andromedae. Much fainter is the 'garnet-star' μ Cephei; its colour is best appreciated by using a pair of binoculars and comparing to the neighbouring α Cephei. These colours are an indication for the surface temperature of the stars.

Procedure

1S. Select an interesting constellation, at not too high an altitude. (Say e.g. Orion.) Sit down in an easy position, illuminate your sketchbook with faint red light.

2S. Draw on a rather big scale. First indicate the brightest stars, paying attention to the proportions and to the direction of the vertical. Use 'alignments'. Then add the fainter stars.

Try to record at least inside a limited area *all* the stars which you are able to observe.

Don't draw open circles, just make black dots. Their size should suggest the brightness of the stars.

3S. Put in your drawing: the direction of the horizon, the position of neighbouring planets, date and hour.

4L. Compare your drawing to a *Star Atlas*. Add in tiny characters the Greek letters or numbers, which are in use to designate them.

5S. Compare the atlas with the sky and try to observe still fainter stars.

(6S.) Draw again the area which you had completely mapped, but now use a field-glass. (a) Focus first for one eye with the fixed eye-piece; (b) then regulate the adjustable eye-piece for the other eye; (c) adjust the distance between both oculars.

Notice how many more stars appear.

(7S.) Look at the same field through your experimental telescope.

(8S.) Select a few stars of different brightness; look up their magnitude in the *Catalogue of Bright Stars*, list them. Try to remember approximately what the stars of these different magnitudes look like; estimate magnitudes for other stars and check them.

References

Bečvář, A.: 1958, *Atlas Coeli*; 1964, *Catalogue*, Praha.
Schürig, R. and Götz, P.: 1960, *Himmelsatlas* (*Tabulae Caelestes*), Mannheim.

Preparation

For each pair: *Norton's Star Atlas*; lamp, dimmed, red; experimental telescope.

A few copies of Schlesinger's *Catalogue of Bright Stars*; a list of star colours is found in: 1921, R. Henseling, *Astronomisches Handbuch*, Stuttgart, p. 206.

A few field-glasses.

Note. This is partly a repetition of exercise A1, it is intended for students who have not worked on the Planetary System.

B16. THE APPARENT MAGNITUDES OF THE STARS (S)

Introduction

The brightness of the stars, as perceived by the eye, is recorded on a scale of magnitudes. A difference of one magnitude corresponds to a brightness ratio K. The constant K is so chosen, that a difference of 5 magnitudes corresponds to a factor of 100.

Thus $100 = K^5$, $\log K = 0.40$ and $K = 2.51$.

GENERAL INSTRUCTION

In order to find the coordinates and the brightness of a relatively bright star we always follow a standard procedure:

(a) find the star in the *Star-Atlas*;

(b) read the right ascension;

(c) consult a star catalogue where the stars are listed in the order of right ascension;

(d) check whether the declination and the magnitude are in reasonable agreement with the atlas.

Excellent catalogues of the brightest stars are found in:

Astronomical Ephemeris ('Mean Places of Stars')

Connaissance des Temps

F. Schlesinger: *Yale Catalogue of Bright Stars.*

Observations with the Naked Eye

1L. Make a list of some stars, now visible and easy to find, so that you have a scale of brightness values, between 0^m and 5^m. Choose these stars, if possible, in one and the same region of the sky.

2. With respect to these standards, estimate the brightness of 6 other stars and record your estimates.

3. Look closely at a faint star. Keep your eyes fixed on it. Is it now more plainly visible or less visible? What are the faintest stars which you are able to perceive *when applying this manner of looking*? Compare with an earlier result: you will be surprised at the importance of this effect!

4. What is the smallest magnitude difference which you are able to perceive? Compare α and β UMa... $\Delta m = 0^m.6$

ε and η UMa... $\Delta m = 0^m.2$

In what direction is the difference?

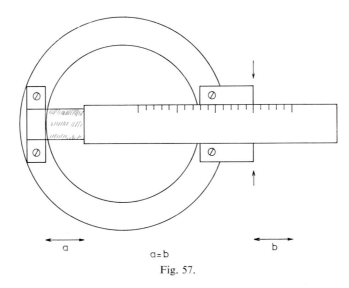

a

a = b

b

Fig. 57.

Observations with a Simple Photometric Device (Figure 57)

In order to compare the brightness of two stars, we insert before the objective of our telescope a small glass prism with a very small refracting angle (about 20′). This prism is covered by a sliding metal strip; on a graduation we read the length of the prism which is free and transmits the light.

The rays of a star will for the greatest part pass on both sides of the strip and will form the stellar image in the normal way. A small part, however, falls on the prism and forms a fainter image, displaced over about 10′. The intensity ratio of the two images is determined by the ratio of the areas, contributing to each of the images.

The area contributing to the main image is, in our case, 6.3 cm². (Make a gross estimate and verify that this is about true.) The effective free area of the prism is b, given on the graduation, multiplied by the width $w = 1$ cm. The ratio therefore is $6.3/b$, corresponding to a magnitude difference of $2.5 \log(6.3/b)$.

5. First push the metal strip sidewards, so that no light is transmitted by the prism. Point your telescope to Mizar (ζ UMa) and Alcor. Now slowly slide the metal strip further and further, uncovering step by step the little prism. – You see the accessory images appearing in a most delightful way. By moving the slide you are able to modify their brightness, and you should try *to make the accessory image of Mizar equal to the main image of Alcor.* Always rotate the photometric head before the objective in such a way that the accessory image comes in the neighbourhoud of the primary image to which it should be compared.

In trying to reach equality, you will soon notice the difficulty of visual stellar photometry. Adjust the focus closely: as soon as this is no longer perfect the main image becomes double, since it is formed by two bundles coming from both sides of the objective. The accessory image and the main image which you want to compare have not quite the same aspect: the first is sharper than the second, because actually the prism acts as a diaphragm which suppresses some aberrations. Fortunately in this case they have about the same colour (spectral classes A2 and A5). You will find that even the direction of your line of sight has some influence.

Adjust the prism so that there can be no doubt that the accessory image is too bright; then shift it till you are sure that now it is too faint. Finally find a position where the balance is as perfect as possible. Look quietly and carefully, but not *too* long in order not to strain the eye.

Make several independent measurements, always accepting them as they are and not trying to correct them afterwards. Take the mean and compute the magnitude difference between the two stars. Compare with the catalogue ($\Delta m = 1^m.4$).

(6.) Similar measurements can be made in the group of the Pleiades. The insertion of the prism doubles the number of stars in the field and increases the beauty of this cluster! – Compare η Tau and 27 Tau

27 Tau and 28 Tau.

(Identify the stars by the insert on Norton's map 5.)

Avoid comparisons between stars differing by more than 2^m.

(7.) From your measurements, find the mean error of your photometric estimates and express it into magnitudes. Even with excellent instruments and experienced observers this error is of the order of $0^m.1$.

(8.) Adjust the prism so as to obtain a magnitude difference of 1^m. Try to remember the impression of two stellar images having this ratio.

9. Derive from your measurements the number K, giving the intensity ratio corresponding to one magnitude.

TABULATION

Name of the star	α	δ	m from catalogue	m estimated	difference
....	....	....	....		
....	....	....	....		
....	....	....	....		
....	....	....	....	....	....
....	....	....	....	....	....
....	....	....	....	....	....

Comparison Mizar – Alcor.

b = free area of prism.

....
....
....

$\bar{b}$ = ...

Δm = ... Δm from catalogue = ...

References

BECKER, W.: 1950, *Sterne und Sternsysteme*, Dresden und Leipzig, § AI2.
SAUER, M. and STRASSL, H.: 1957, *Veröffentl. Sternw. Bonn* **46,** pp. 10 and 11.
Nomograms in LANDOLT-BÖRNSTEIN, p. 319.

Preparation

For each pair: experimental telescope with photometric head (see p. 166); *Astronomical Ephemeris*; flashlight; *Star Atlas.*

B17. THE DIAMETER OF STELLAR IMAGES
AS A MEASURE FOR THE BRIGHTNESS (L)
(Structure of a Stellar Image)

The Problem

Look at a photograph of the open cluster Praesepe (= M44), in the constellation Cancer. Notice how different are the sizes of the stellar images. These differences have nothing to do with the real diameters of the stars, but they are related to the apparent magnitudes of the stars, which vary between 6^m and 13^m. Due to instrumental aberrations and (mainly) to atmospheric scintillation, the image of a star is always somewhat blurred. On the photographic plate the sensitivity threshold i_0 is reached quite near to the central maximum for weak stars, while for bright stars it is exceeded over a great part of the image. The diameters of the images may be used for a rough determination of the brightness. At the same time we shall find the distribution of light in a stellar image.

Procedure

1. Compare your photograph with the map, mentioned at the end of the exercise, and showing the photographic magnitudes. Identify some of the stars.

2. Measure to 0.1 or 0.05 mm the diameter of a number of stellar images by means of a scale and a magnifier. In order to eliminate possible astigmatism, we measure each image twice, in directions perpendicular to each other, and we take means. Give special care to the weakest stars.

3. We look now for an empirical relation between the diameter D and the apparent magnitude m. This we do graphically. But in many cases it is easier to have also a simple interpolation formula. Try to adapt to your graph the following formulae successively:

$$m = a - bD$$

$$m = a - b\sqrt{D}$$

$$m = a - b\log D .$$

4. Which formula gives the best fit?

Estimate the remaining deviations in magnitudes. How could you explain that each plate is found to require its own formula?

5. From the function found, derive the light distribution in each stellar image $i = If(r)$, where I is the central intensity, i the intensity at a distance r from the centre (Figure 58).

The treshold intensity i_0 is reached at a radius r_0, where $i_0 = If(r_0) =$ constant.

If we reduce the magnitudes to intensities, the function which we found empirically may be written: $I = \varphi(r_0)$, since the central intensities are proportional to the total intensities. Consequently

$$f(r_0) = \frac{\text{constant}}{\varphi(r_0)}$$

and generally

$$f(r) = \frac{\text{constant}}{\varphi(r)}.$$

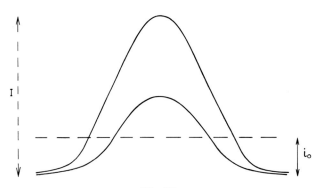

Fig. 58.

Draw the intensity distribution in the stellar images of your photograph.

References

LUNDMARK, K.: 1932, in *Handbuch der Astrophysik*, Vol. V¹, p. 296.
STOCK, J. and WILLIAMS, D.: 1962, in *Stars and Stellar Systems*, Vol. II, Chicago, p. 393.

Preparation

For each pair: a photograph of the cluster Praesepe, copied on glass; a reproduction on paper of Hertzsprung's survey map (1917, *Astron. Nachr.* **203,** 268). – Photographic reproduction on glass of a wedge-scale for the measurement of minute objects (p. XXIII); magnifier; two sheets of rectangular graph paper.

This exercise gives the students some insight in the structure of a photographic stellar image. The argument by which the distribution of light is obtained looks simple but is not always easily grasped.

B18. PHOTOELECTRIC PHOTOMETRY OF
STELLAR IMAGES (L)

The Problem

This is the standard method of comparison by photography of the brightness of stellar images. We insert each of them in a narrow beam of light and measure the resulting reduction in light intensity. It is not our purpose to measure the absorption point by point in the image; we want an *integrated* effect. It is clear that an increase in the brightness of the star will result in an increased integrated absorption. The relation between both variables cannot be predicted theoretically and should be determined empirically for each plate.

Procedure

1. You will receive a record of the constellation Orion, photographed on an ordinary blue-sensitive plate. Compare this photograph to the *Star Atlas*, identify some characteristic stars.

2. Put the plate on the stage of the microphotometer, gelatine-side down. Never press the plate down, for fear of scratches. Keep it gently in contact with the paper which covers the stage. Switch on your microphotometer. (For a description see p. XVIII and p. 99.) Move the lower lens slightly till the deflection of the micro-amperemeter is maximum. Read first the deflection when a part of the plate without stellar images is put on the aperture. Repeat this measurement for two or three other parts of the clear plate. Notice how the slightest fog on the plate immediately affects the reading. There may also be fluctuations due to the varying tension in the mains.

3. Now insert a rather strong, well-identified and single stellar image. It must coincide *as precisely as possible* with the centre of the aperture, at the intersection of the OX and OY lines on the stage. This is not so easy, since you are obliged to look from a slanting direction, say along the OY line. You will easily manage to set the star on OY; by a slight motion towards or away from you, try to find the position where the μ-amperemeter gives the smallest current and where, consequently, the integrated absorption is a maximum. This should be repeated several times till the best position is well ascertained. It will help if your partner follows the needle and catches the precise minimum.

4. Repeat this measurement for a number of stars of very different magnitude. Use only well-identified and single stellar images, designated by a number or by a letter in the atlas. Fainter stars are identified by comparison with Bečvář, *Atlas Coelestis*.

5. Look up the *photographic* magnitudes in Schlesinger's *Catalogue of Bright Stars*; the latest (3rd) edition gives the visual magnitudes and the colour index. Fainter stars may be found in the *Bonner Durchmusterung* charts, combined with

the catalogue of the *Astronomische Gesellschaft*, but the identification will be difficult.

6. Plot the deflection of the micro-amperemeter against the stellar magnitude. If some point deviates manifestly from the main curve, check the identification! Ascertain the quality of the images with a magnifier and check whether they have perhaps a faint companion or whether they are elongated (border of the field!).

7. Estimate the precision of your measurements in fractions of a stellar magnitude.

Reference

STOCK, J. and WILLIAMS, D.: 1962, in *Stars and Stellar Systems*, Vol. II, p. 394.

Preparation

For each pair: microphotometer, connected to the contact box for 12 V or 24 V; photograph of an easily recognizable star field (Pleiades, Orion); a stencilled list of photographic magnitudes for easily identifiable stars brighter than 6^m, in the same field; *Norton's Star Atlas*; a few copies of Bečvář, *Atlas Coeli* 1950.00 for this part of the sky.

The exposure of the plate, the little hole in the stage, the temperature of the microphotometer lamp and the sensitivity curve of the photovoltaic cell must be adapted to each other, so that the calibration curve keeps a sufficient slope over the whole range of stellar images. This is easily attained.

B19. THE ATMOSPHERIC EXTINCTION (S)

The Principle

By absorption and scattering in the terrestrial atmosphere, the light of a star is weakened according to the simple exponential law $i = i_0 e^{-k \sec z}$, where z represents the zenith distance. The coefficient k may be determined by comparing stars at high and at low altitudes.

PROCEDURE (ON CLEAR NIGHTS)

1L. At altitudes above 60° all stars show practically the same extinction. We shall call them 'zenith stars'. By means of the *Star Atlas*, select about 10 such stars of very different brightness, e.g. in the constellation of Auriga; to these all other, lower stars will be compared.

2S. Observe a star A at low altitude h, and look for a zenith star in your list which seems to have precisely the same brightness. If no such star can be found, interpolate between two zenith stars z_1 and z_2. Identify the star A with the atlas; determine roughly its height h with the cross-staff.

3S. Repeat this observation for a few other stars at low altitudes. Try to find at least one in the altitude range of 3°–5°.

4L. Look up the brightness of the zenith stars used in the *Catalogue of Bright Stars* or in the *Astronomical Ephemeris*.

With respect to them the observed brightness of the low stars will be found to be less than mentioned in the catalogue. Put the difference equal to $\Delta m = m - m'$, where m' applies to zenith stars.

5L. We calculate now the extinction coefficient.

For a low star:
$$\log_{10} i = \log_{10} i_0 - 0.434 \, k \cdot \sec z$$
$$2.5 \log_{10} i = 2.5 \log_{10} i_0 - 1.08 \, k \cdot \sec z.$$
$$m = m_0 + 1.08 \, k \cdot \sec z.$$

For a zenith star:
$$m' = m_0' + 1.08 \, k.$$

TABULATION

star	h	$\cos h$	$\dfrac{1}{\cos h} = \sec z$	m obs.	m catal.	Δm
z_1					–	
z_2					–	
(A)	–	–	–	–	–	–

If both look equally bright: $\Delta m = m_0' - m_0 = 1.08 \, k \, (\sec z - 1)$. Plot Δm against $\sec z$ and draw the best straight line between the observed points. From the slope derive k. (Compare the graph in exercise B2.)

6L. In what proportion (expressed in magnitudes) has the light of a zenith star been weakened by our atmosphere?

References

ALLEN, C. W.: 1963, *Astrophysical Quantities*, London, p. 122.
HARDIE, R. H.: 1962, in *Stars and Stellar Systems*, Vol. II, Chicago, p. 184.
LUNDMARK, K.: 1932, in *Handbuch der Astrophysik*, Vol. I, Berlin, p. 566.

B20. STELLAR SPECTRA (L)

We study reproductions of spectra, recorded at the Michigan Observatory. Handle these spectra carefully. Orient the violet side to the left, the red side to the right. Where necessary, use a magnifier; it has a plane-convex lens: turn the convex side towards your eye, bring the eye quite close to it.

Procedure

1. Look at the spectrogram. It is a positive print which has been artificially widened, and in which the dark Fraunhofer lines appear. Why does this strip not extend further towards the left and the right? You have also received a scale on which wavelengths are marked (Figure 59). Bring the spectrum and the scale into coincidence.

2. Compare the spectrum with the photographs of characteristic spectral types in books, and get first a general impression of the class. Roughly speaking, the weaker the spectrum in the violet and the more spectral lines present, the later is the spectral type. Compare the relative strengths of the hydrogen lines, the helium lines, the metal lines and the band spectra (consult the determination table given at the end of this instruction).

3. Now read the detailed description of spectral types in the neighbourhood of that which you have guessed, and use their characteristics for a sharper differentiation.

Compare especially the strength of neighbouring lines of very different excitation and ionization.

Estimate the strength of some characteristic lines in the scale: 3 = strong, 2 = rather strong, 1 = faint, 0 = lacking.

Finally give your judgment about the spectrum.

4. Exchange your plate for others, showing quite different spectral types, and study several of them.

5. The ordinary stellar classification is mainly determined by the surface temperature of the stars.

About 1945 Morgan and Keenan introduced another dimension: *luminosity*, which is directly connected with the *density* (or the gravitational potential). By careful study of minute details it proved possible to distinguish, within the same type or sub-type, stars of different luminosity classes; they are labelled by adding the roman numerals Ia, Ib, II, ... V, the first one corresponding to super-giants, the last one to dwarfs.

It is interesting to apply these criteria to some of your spectra. Study especially ε Ori, where the characteristics are pronounced and clear. We use among others the lines 4089 (Si IV), 4009 (He I), 4144 (He I).

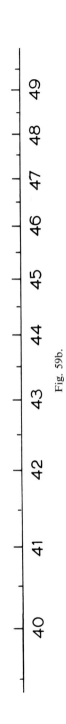

Fig. 59b.

Fig. 59a.

4009 < 4089	4009 = 4089	4009 > 4089
4072 < 4089	4072 = 4089	4072 > 4089
4144 < 4119	4144 = 4119	4144 > 4119
Luminosity I	Luminosity III	Luminosity V

For δ Gem we compare with 4226 (Ca I resonance line).

4172 > 4226	4172 = 4226	4172 < 4226
Luminosity Ia	Luminosity III	Luminosity V

(6.) It is also instructive to examine low dispersion spectra as obtained with an objective prism. Or high-dispersion coudé spectra.

Determination table

$$
\begin{array}{l}
\text{H lines clearly the strongest ones}
\begin{cases}
\text{He or He}^{+}\text{ visible} \dotfill \text{B} \\[2ex]
\text{He or He}^{+}\text{ absent}
\begin{cases}
\text{H}_{\varepsilon}+\text{Ca II } 3969 \gg \text{Ca II } 3934 \\
\text{(Ca 4226 nearly invisible,} \\
\text{G 4310 lacking, Mg II 4481 distinct)} \dotfill \text{A} \\
\text{3969 and 3934 of the same order} \\
\text{(Ca 4226 faintly visible,} \\
\text{G 4310 mostly visible)} \dotfill \text{F}
\end{cases}
\end{cases} \\[12ex]
\text{H lines not the strongest}
\begin{cases}
\text{Ca 4226 of the same order as H}_{\gamma}\text{ and H}_{\beta} \dotfill \text{G} \\[1ex]
\text{Ca 4226 > H}_{\gamma}\text{ and H}_{\delta}
\begin{cases}
\text{no clear molecular bands} \dotfill \text{K} \\
\text{clear molecular} \\
\text{bands (4582, 4762)}
\end{cases} \dotfill \text{M}
\end{cases}
\end{array}
$$

Some characteristic lines

H	He I	He II	Ca I	Ca II	Mg II
η 3835	4026	3924	4226	3934	4481
ξ 3889	4472	4026		3969	
ε 3970		4200			
δ 4102		4542			
γ 4340		4686			
β 4861					

References

'The Henry Draper Catalogue': 1918–1924, *Harvard Annals*, **91–99**. Especially **91**, pp. 5–11.

The detailed description of the spectral types is found reproduced in many books, e.g. F. J. M. Stratton: 1924, *Astronomical Physics*, London, p. 110. – H. Curtis: 1932, in *Handbuch der Astrophysik*. V¹, Berlin, p. 44. – *Trans. Intern. Astron. Union*: 1935, Vol. V, p. 181. – LANDOLT-BÖRNSTEIN, p. 285.

The detailed subdivision of the spectral types and the luminosity classes is found in the highly valuable *Atlas of Stellar Spectra*, by Morgan, Keenan and Kellman (Chicago, 1943). Its use is facilitated by comparing with the monograph of Ch. Fehrenbach 1958, in *Handbuch der Physik*, **50**, p. 44, where one sees at a glance the lines to be compared, but where the reproductions are not satisfactory.

Preparation

Reproductions from Rufus: 1923, *Publ. Michigan*, p. 258–259, from which the spectral type indication has been removed; these reproductions are interchanged in the course of the evening. Descriptions of the Harvard types.

A few reproductions from the Morgan-Keenan-Kellman *Atlas* and the corresponding figures from Fehrenbach's monograph. Prints of each separate Atlas plate may be ordered from the Yerkes Observatory.

A few reproductions of objective-prism spectra; see O. Gingerich: 1964, *Sky and Telescope* **28**, p. 80.

Reproductions of high-dispersion coudé spectra (1 Å/mm) of the Mt. Wilson and Palomar Observatories are for sale at the Calif. Inst. of Techn., Pasadena (Nr. 304); the series of 15 spectra includes also a few examples of the differences between giants and dwarfs.

Contact prints should also be shown of authentic, not magnified, not widened stellar spectra, as they have been available for sale at the Yerkes Observatory. For these a pulpit (p. XXII) and a magnifier will be necessary.

The following recent publication is also very suitable: *An Atlas of Low-Dispersion Grating Stellar Spectra*, 1968, Kitt Peak National Observatory, Tucson, Arizona.

B21. THE COLOURS OF THE STARS (L)

The Principle

In the introduction of exercise B15 we have already noticed the colour differences between the stars. For scientific work these colours are studied quantitatively by photography (or by photoelectric measurements). According to whether we use blue-sensitive plates or orthochromatic, yellow-sensitive plates with a yellow filter, we obtain *photographic* or *photovisual* magnitudes. Their effective wavelengths correspond approximately to $\lambda = 4400$ Å and to $\lambda = 5450$ Å. The higher the temperature of the star, the greater the ratio i (blue)/i (yellow) and the smaller the *colour index* $c = m_{pg} - m_{pv}$. – The colour index, thus, is a measure for the 'yellowness' of the star and for its temperature. – We study some photographs of binaries, whose components have very different temperatures. An *objective grating* has been placed before the objective, giving two diffraction images for each star, their magnitude being about $1^m.0$ lower than the main image. By shifting the plate a little, a number of consecutive exposures have been obtained. – As to the function of the objective grating, cf. exercise A19, § 12.

Procedure

1. Compare first cursorily the photographic and the photovisual plate, using a magnifier and note the striking difference. – Always handle photographic plates with care! Do not touch the surfaces, handle them by the edges. When the gelatine layer is turned towards you, the characters written on the plate are seen in the right way.

2. Estimate the magnitude difference between the two components: (a) on the photographic record, (b) on the photovisual record. In doing this, rely on the magnitude interval, as given by the objective grating; compare also the central image of the weak star with the first diffraction image of the brighter star. – Do not confuse the signs of the magnitude differences! Always take the differences in the same sense, both for the photographic and for the photovisual records. Note that m *increases* if the star is *less bright*.

List directly the spectral types of the stars, in the same order as that in which you have estimated the magnitudes.

Let each of the two students make an independent estimate. Compare and take the average.

3. You can now determine the colour difference of the two stars. For:

$$(m_{pg} - m'_{pg}) - (m_{pv} - m'_{pv}) = (m_{pg} - m_{pv}) - (m'_{pg} - m'_{pv}) = c - c'.$$

Compute this for all available plates.

4. Of the stars photographed here we know the spectral types. In the order of

decreasing temperature these types are:

$$B_0, B_1, ..., A_0, A_1, ..., F_0, F_1, ..., G_0, G_1, ..., K_0, K_1, ..., M_0, M_1, ...$$

We ask, how does the colour index depend on the temperature? Take as an abscissa the spectral type, as an ordinate the colour index. Remember that the colour index, by convention, is put equal to zero for an A_0 star; plot therefore first the colour indices for the pair $A_0 - G\ 7$. For the other binaries you know only *the difference* of the colour indices; plot each of these pairs at such a height that the curve runs fluently through all points observed.

(5.) In order to *understand* this graph, we substitute for the spectral types the corresponding temperatures T, or better $1/T$ (Table in B22). If the stars were shining as black bodies and if our colour filters were monochromatic, elementary theory shows that the colour index would be approximately of the form: $c = a + b(1/T)$. – See how far this proves true.

References

ALLEN, C. W.: 1963, *Astrophysical Quantities*, London, p. 201.
JOHNSON, H. L. and MORGAN, W. W.: 1953, *Astrophys. J.* **117**, 350.
LANDOLT-BÖRNSTEIN, p. 298.

TABULATION

Star	$m_{pg} - m'_{pg}$	$m_{pv} - m'_{pv}$	$c - c'$	Spectra
....	....	....	....	....
....	....	....	....	....

Preparation

Photographs of 7 binaries, taken with an objective grating (a) on a blue-sensitive emulsion; (b) on a yellow-sensitive emulsion with filter. Positive contact prints of each of these 7×2 photographs are distributed and now and then interchanged. Each print on glass has 2 cm $\times$ 2 cm.

For each pair: pulpit for illumination of the glass prints; magnifier; rectangular coordinate paper.

B22. THE UBV PHOTOMETRIC SYSTEM (L)

Introduction

In order to characterize the colour of a star by *one* single number we have used the *colour index*: $m_{pg} - m_{pv}$. This would indeed be sufficient, if the energy distribution in stellar spectra corresponded to that of black bodies. The observation shows, however, that there are considerable deviations, due to the accumulation of Fraunhofer lines, to interstellar reddening, or to differences in composition.

The shape of the energy curve is better defined, if we are allowed to quote *two* indices. After careful consideration Johnson and Morgan introduced for that purpose the UBV system, which is now widely used. Photoelectric measures of the stellar radiation are made through 3 filters, of which the effective wavelengths correspond roughly to λ 3650 (U), λ 4400 (B), λ 5500 (V = visual). These measures are expressed in magnitudes, and the two indices $m_U - m_B$, $m_B - m_V$ are used as characteristics for a given spectrum.

Purpose / Material

We intend to study how the indices U-B, B-V must be expected to vary with the spectral class along the main sequence. Our basic material (Figure 60) derives from photo-electric scans, representing the intensity of stellar radiation as a function of wavelength. The scans are made with a wide slit, they do not represent the true continuum, but include the effect of accumulated Fraunhofer lines. The intensity is measured per unit of wavelength interval $\Delta\lambda$ and expressed in magnitudes in the usual way. Increasing magnitudes correspond to decreasing intensities. We have selected stars from the luminosity class V and we have avoided stars with galactic latitudes between $-10°$ and $10°$, in order to avoid the worst interstellar reddening.

The stars selected are: 10 Lac, η UMa, α Lyr, β Ari, ρ Gem, π^3 Ori, λ Ser, 51 Peg, ε Eri, HD 202560.

Procedure

1. Look at these curves, compare them to black body curves and notice the great deviations, of which the causes are easily understandable.

2. Let now photoelectric measurements of these stars be made through filters around $\lambda\lambda$ 3650, 4400, 5400. Actually such a filter transmits radiation within a rather wide band of the spectrum and the effect should be computed by an integration. But in order to save time we shall assume that each filter is strictly monochromatic.

Read for each spectrum the U, B, V magnitudes. Where it is necessary, extrapolate the curves.

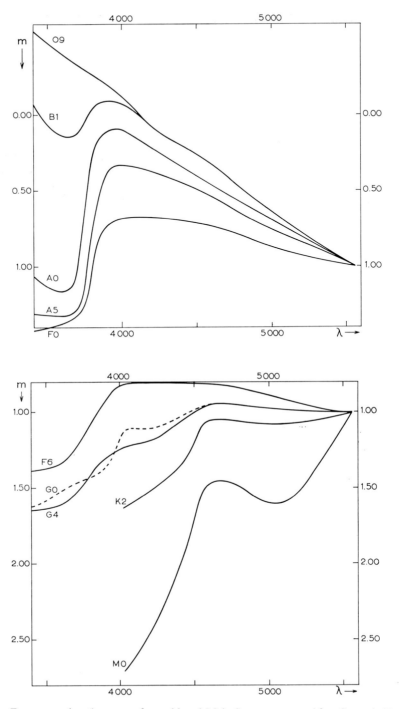

Fig. 60. Energy-wavelength curves of unreddened Main-Sequence stars. After CODE, A. D.: 1960, in *Stars and Stellar Systems*, Vol. VI, p. 83. Modified after MELBOURNE, W. G.: 1960, *Astrophys. J.* **132**, 101; complemented after WILLSTROP, R. V.: 1965, *Mem. Roy. Astron. Soc.* **69**, 83.

3. Compute for each star the indices U-B, B-V.

4. Reduce them to the conventional values U-B=0, B-V=0 for A0 V stars. – (Realize why a constant number may always be added to a system of colour indices.)

5. Plot the indices in a diagram and trace a smooth curve between the points. In conformity with the general practice, let the algebraic values of the indices increase in the downward direction.

6. Compare with similar curves found in the literature, and explain why some differences may be expected.

(7.) Plot the U-B, B-V indices for stars of one selected spectral type, for example B0.5-V, reddened to different degrees by interstellar matter. Much material is found in the *Catalogue of Bright Stars* (3rd ed.), constellations Cygnus, Sagittarius, Monoceros; and in Hiltner's Catalogue, 1956, *Astrophys. J.*, Suppl. II, 389. – Trace *the reddening path*.

References

JOHNSON, H. L.: 1963, in *Stars and Stellar Systems*, Vol. III, p. 204.
JOHNSON, H. L. and MORGAN, W. W.: 1953, *Astrophys. J.* **17**, 313. See also the data for clusters (B33).
JOHNSON, H. L.: 1955, *Ann. Astrophys.* **18**, 292.

Preparation

For each student: coordinate paper; measuring scale; slide rule.

A few copies of the *Catalogue of Bright Stars* and of HILTNER's *Catalogue*.

Note. – Another set of data on the smoothed energy distribution in the spectra of stars is found in LANDOLT-BÖRNSTEIN, p. 388 (after Lamla, corrected).

B23. INTERSTELLAR EXTINCTION (L)

The Problem

Many observations became understandable only when astronomers realized that there are huge clouds of gas and dust in interstellar space, by which the light of far-off stars is dimmed and reddened. In order to reduce our observations for this effect, we have to find how this extinction depends on wavelength; our results will also help to understand the nature of the dust.

We shall look for two stars, of which the spectral lines very perfectly match each other, while however one of them is considerably reddened with respect to the other.* Comparing the continuum brightness in a number of points, we find how the extinction depends on the wavelength. Most suitable for this purpose are O and B stars, which have few spectral lines. For the first star, let the brightness $i_{0\lambda}$ at a wavelength λ be dimmed by interstellar extinction; we observe $i_\lambda = i_{0\lambda} \times 10^{-A}$. Thus:**

$$\log i = \log i_0 - A = \log i_0 - \kappa_\lambda M,$$

where κ_λ is the extinction coefficient per gram, M the total mass of interstellar matter along the line of sight.

For the other star we have similarly:

$$\log i' = \log i_0' - A' = \log i_0' - \kappa_\lambda M'.$$

The ratio of two brightnesses is given by:

$$\log i' - \log i = \log i_0' - \log i_0 + \kappa_\lambda (M - M') = C + \kappa_\lambda (M - M').$$

But for the constants C and $(M - M')$, we find the extinction law $\kappa(\lambda)$.

Procedure

By a very careful spectrographic investigation, Miss Divan (1954) found several pairs of stars, showing precisely the same spectral type and luminosity class, but a very different reddening. From these we select two pairs and list the values of $\log i' - \log i$, found by Miss Divan for a number of selected wavelengths (Table I). These are actually two determinations of $\kappa(\lambda)$, but for the constants.

1. Let us check whether the two functions $\kappa(\lambda)$ agree. Plot $(\log i' - \log i)_1$ against $(\log i' - \log i)_2$.

2. Do you find a linear relation between the two quantities? That would mean: a linear relation between the two quantities $\begin{cases} C_1 + \kappa_\lambda (M - M')_1 \\ C_2 + \kappa_\lambda (M - M')_2 \end{cases}$

---

* For small differences in the spectra, a correction may be introduced.
** In the following formulae, the symbols i, i_0, i', i_0', A, all refer to monochromatic radiation and are functions of the wavelength λ; λ being the same inside each equation.

TABLE I (Divan, 1954)

λ	$1/\lambda$	$\log i' - \log i$	
0.313	3.20	+0.642	+0.345
0.322	3.11	+0.571	+0.298
0.333	3.00	+0.499	+0.245
0.365	2.74	+0.374	+0.105
0.402	2.49	+0.204	−0.018
0.443	2.26	+0.054	−0.142
0.488	2.05	−0.076	−0.257
0.562	1.76	−0.343	−0.428
0.611	1.64	−0.428	−0.552
		$\begin{cases} HD\ 188.209 \\ HD\ 195.592 \end{cases}$	$\begin{cases} S\ \text{Mon} \\ HD\ 205.196 \end{cases}$

TABLE II (Whitford, 1948)

λ	$1/\lambda$	$\log i' - \log i$
3200	3.12	+0.550*
3530	2.83	+0.502*
4220	2.37	+0.400
4880	2.05	+0.324
5700	1.75	+0.256
7190	1.39	+0.140
10300	0.97	+0.000
20800	0.48	−0.056

* Corrected after Divan (1954, p. 520)

What do you conclude about $\kappa_{\lambda 1}$ and $\kappa_{\lambda 2}$? And about the ratio $(M - M')_1/(M - M')_2$?

Find the best approximation for this ratio.

3. In order to combine the results from both pairs, plot $(\log i' - \log i)_1$ and $(\log i' - \log i)_2 (M - M')_1/(M - M')_2$ against $1/\lambda$. (We prefer to take as an abscissa $1/\lambda$ because then the plot is more nearly linear.) If the two functions $\kappa(1/\lambda)$ are the same, the two plots can be combined by shifting them in the direction of the ordinates. Take means and find $\kappa(1/\lambda)$, but for an unknown *constant of proportionality* and an unknown *additional constant*.

4. We want to get rid of the unknown additional constant. This can be done if we are able to extend our curve to $\lambda \to \infty$ or $(1/\lambda) \to 0$, where evidently the extinction disappears.

Let us make use of the photoelectric observations of Whitford (1948, *Astrophys. J.* **107,** 48), who compared ε Per and φ Per and extended his observations from 3200 to 21'000 Å. Compare his results to yours, along the lines, followed in (3) above. Find the function $\kappa(1/\lambda)$ and extrapolate to $\lambda \to \infty$. (For this extrapolation, you might use as a guide Rayleigh's $(1/\lambda)^4$ law, which must apply for the longest wavelengths.)

5. The zero-point being now determined, you are able to make a new graduation for the graph $\kappa(1/\lambda)$. Many authors find κ proportional to $1/\lambda$ in the photographic and visual region. Is this confirmed?

6. You are now able to find the very important ratio between the total visual extinction A_V and the *colour excess* E_{B-V}. From your curve find A_V/E_{V-B}, the representative wavelengths being $\lambda\lambda$ 4350 and 5430 Å.

7. We demonstrate the use of this quantity.

Consider the star Nr. 229 in Hiltner's catalogue (1956):

Spectral class $B1-Ia$, $B-V = +1.20$, $U-B = -0.01$, $m = 4.23$.

For such stars, unreddened: $(B-V)_0 = -0.26$, $M = -6.5$.

M, m, A refer to the visual region.

Compute the distance modulus, not taking account of interstellar absorption.

Then take this effect into account: compute E_{B-V}, A, $m-A$, and the correct distance modulus. This star has a galactic latitude of $-0°.1$.

Note. In this exercise, we have assumed that the extinction in interstellar space is the same function of wavelength in all directions. This is in general true, but there are exceptions.

Reference

SHARPLESS, S.: 1963, in *Stars and Stellar Systems*, III, p. 225.

Preparation

For each student: rectangular coordinate paper; slide rule.

Many more data are found in: Divan, 1954, *Ann. Astrophys.* **17**, 456. We have picked out only a few at random in order to save time.

B24. THE NEAREST STARS (L)

The Problem

Adjoined to this text is the raw material: a list of some 40 stars with the greatest known parallaxes, for which some properties have been measured. Of these well-known stars we shall compute several other properties and then survey the whole collection. Each pair of students works on 5 stars; all results are finally combined.

Procedure

1. Compute for each star the distance in parsec $D = 1/p$, where p is expressed, as usual, in seconds of arc.

2. See whether the apparent magnitude of the 50 stars is more or less a measure for their distance.

3. Compute for each star the absolute magnitude

$$M = m + 5 + 5 \log p.$$

4. For our further calculations the visual magnitude M_v must be first reduced to the *bolometric magnitude*, which represents the total amount of radiation emitted. The necessary *bolometric correction* $(B.C.)$ is listed in a special table below. (Try to understand for yourself why the correction depends in such a peculiar way on the temperature.) Thus compute:

$$M_{bol} = M_v - (M_v - M_{bol}) = M_v - B.C.$$

5. Write after each spectral class the corresponding effective temperature (see Table I). Then compute the diameters of these stars, expressed in the solar diameter. Use the expressions:

$$I_{bol} = C \cdot 4\pi R^2 \cdot \sigma T_e^4 \quad \text{and} \quad \left(\frac{I}{I_0}\right)_{bol} = \left(\frac{R}{R_0}\right)^2 \left(\frac{T_{e*}}{T_{e0}}\right)^4.$$

Use logarithms and magnitudes.

6. Plot for these 50 nearby stars the Hertzsprung-Russell diagram. As an abscissa, do not take spectral types but $\log T_e$ (Table II). In how far is your diagram different from the 'ordinary' figure, which you find in the textbooks? Note some striking properties of this star collection.

How are the diameters distributed over the diagram? Could you draw lines of equal diameter?

(7.) What is (roughly) the mean distance of two neighbouring stars? How many percent are members of double or multiple systems? How many percent are white dwarfs?

(8.) Plot the luminosity function $\varphi(M)$ for these stars. Count the number of stars between $M_v - \frac{1}{2}$ and $M_v + \frac{1}{2}$ and reduce this to the number of stars in a pc^3. Compare to the literature.

TABULATION

Star Nr.	$D = \dfrac{1}{p}$	M_v	B.C.	M_{bol}	T	$10 \log T$	$M + 10 \log T$	$\dfrac{M + 10 \log T}{5}$	$\log \dfrac{R}{R_0}$	R
......	....	......	....	......	......	....	....	....	....	....
......	....	......	....	......	......	....	....	....	....	....

References

JENKINS, L. F.: 1952, *General Catalogue of Trigonometric Parallaxes*, with Supplement 1963, New Haven.

See the nomograms in: 1965, LANDOLT-BÖRNSTEIN, VI, Bd. 1, pp. 282–283, and last page on back cover.

STRUVE, O.: 1950, *Stellar Evolution*, Princeton, p. 33 –.

Preparation

For each student, rectangular coordinate paper; logarithms.

TABLE I

Bolometric Correction and Stellar Temperatures
(C. W. ALLEN: 1963, *Astrophysical Quantities*, p. 201.)

Spectrum (Luminosity class V)	B.C.	$T_e(K)$
O5	4.6	35000
B0	3.0	21000
B5	1.6	13500
A0	0.68	9700
A5	0.30	8100
F0	0.10	7200
F5	0.00	6500
G0	0.03	6000
G5	0.10	5400
K0	0.20	4700
K5	0.58	4000
M0	1.20	3300
M5	2.1	2600

The Nearest Stars

(After W. GLIESE: 1963, in Landolt-Börnstein, VI, Bd. 1, 598.)

Nr.	Name	α 1950	δ 1950	p″	μ (″/a)	ϑ*	v_r (km/sec)	m_v	Sp
1	Sun							$-26^m.73$	G 2 V
2 [6]	Proxima Cen	14ʰ 26ᵐ.3	-62°28′	0″.762	3″.85	282°		10.68	M 5 e
	α Cen A	14 36 .2	-60 38	0.751	3.68	281	-25	0 .02	G 2 V
	...B						-21	1 .35	K 5 V
3 [2]	Barnard's Star	17 55 .4	+4 33	0.545	10.34	356	-108	9 .54	M 5 V
4 [6]	Wolf 359	10 54 .1	+7 19	0.427	4.71	235	+13	13 .66	dM 6 e
5 [2]	BD +36° 2147	11 0 .6	+36 18	0.396	4.78	187	-86	7 .47	M 2 V
6	α CMa A	6 42 .9	-16 39	0.375	1.32	204	-8	-1 .47	A 1 V
[4]	...B							8 .67	DA
7	L 726-8A	1 36 .4	-18 13	0.371	3.36	80	+29	12 .45	dM 6 e
[6]	...B							12 .95	dM 6 e
8 [6]	Ross 154	18 46 .7	-23 53	0.340	0.72	104	-4	10 .6	dM 4 e
9	Ross 248	23 39 .4	+43 55	0.316	1.60	176	-81	12 .24	dM 6 e
10	ε Eri	3 30 .6	-9 38	0.303	0.98	271	+15	3 .73	K 2 V
11	Ross 128	11 45 .1	+1 6	0.298	1.40	153	-13	11 .13	dM 5
12	L 789-6	22 35 .8	-15 36	0.298	3.25	45	-60	12 .58	dM 6 e
13 [2]	61 Cyg A	21 4 .7	+38 30	0.292	5.22	52	-64	5 .19	K 5 V
	...B						-64	6 .02	K 7 V
14	α CMi A	7 36 .7	+5 21	0.287	1.25	214	-3	0 .34	F 5 IV-V
[4]	...B							10 .7	DF
15	ε Ind	21 59 .6	-57 0	0.285	4.69	123	-40	4 .73	K 5 V
16 [1]	BD +43° 44 A	0 15 .5	+43 44	0.278	2.90	82	+14	8 .07	M 1 V
	...B						+21	11 .04	M 6 V
17	BD +59° 1915 A	18 42 .2	+59 33	0.278	2.29	325	+1	8 .90	dM 4
	...B						+14	9 .69	dM 5
18	τ Cet	1 41 .7	-16 12	0.275	1.92	297	-16	3 .50	G 8 Vp
19	CD -36° 15 693	23 2 .6	-36 9	0.273	6.90	79	+10	7 .39	M 2 V
20	BD +5° 1668	7 24 .7	+5 23	0.266	3.73	171	+26	9 .82	dM 4
21	CD -39° 14 192	21 14 .3	-39 4	0.255	3.47	251	+21	6 .72	M 0 V
22 [5]	CD -45° 1841	5 9 .7	-45 0	0.251	8.72	131	+242	8 .8	sdM 0

* Position angle of proper motion vector.

Table II (Continued)

Nr.	Name	α 1950	δ 1950	p"	μ ("/a)	ϑ	v_r (km/se$_v$)	m_v	Sp
23³)	Krüger 60 A	22ʰ 26ᵐ.2	+57° 27	0".249	0".87	245°	− 24	9ᵐ.82	M 4 V
⁶)	...B						− 28	11 .4	M 6 Ve
24	Ross 614 A	6 26 .8	− 2 46	0.248	1.00	131	+ 24	11 .2	dM 4 e
	...B							14 .8	(M)
25¹)	BD − 12° 4523	16 27 .5	− 12 32	0.244	1.18	183	− 13	10 .13	dM 4
26⁴)	van Maanen's Star	0 46 .5	+ 5 9	0.236	2.98	155	?	12 .36	DG
27	Wolf 424 A	12 30 .9	+ 9 18	0.228	1.78	280	− 5	12 .7	dM 7 e
	...B							12 .7	dM 7 e
28	BD + 50° 1725	10 8 .3	+ 49 42	0.222	1.45	249	− 27	6 .59	dM 0
29	CD − 37° 15 492	0 2 .5	− 37 36	0.219	6.11	112	+ 24	8 .59	dM 3
30²)⁶)	BD + 20° 2465	10 16 .9	+ 20 7	0.213	0.49	264	+ 10	9 .43	M 4.5 Ve
31	CD − 46° 11 540	17 24 .9	− 46 51	0.213	1.06	147		9 .34	M 4
32	CD − 44° 11 909	17 33 .5	− 44 17	0.209	1.15	217		11 .2	M 5
33	CD − 49° 13 515	21 30 .2	− 49 13	0.209	0.81	184	+ 18	8 .9	M 3
34	BD − 15° 6290	22 50 .6	− 14 31	0.206	1.12	124	+ 9	10 .17	dM 5
35	BD + 68° 946	17 36 .7	+ 68 23	0.205	1.31	196	− 17	9 .15	M 3.5 V
36⁴)	L 145-141	11 43 .0	− 64 34	0.203	2.68	97		11 .47	DC
37	o² Eri A	4 13 .0	− 7 44	0.202	4.08	213	− 42	4 .48	K 1 V
	...B						− 42	9 .50	DA
⁴)	...C						− 45	11 .1	dM 4 e
38	BD + 15° 2620	13 43 .2	+ 15 10	0.202	2.30	129	+ 15	8 .47	M 4 V
39	α Aql	19 48 .3	+ 8 44	0.198	0.66	54	− 26	0 .78	A 7 IV, V
40⁶)	BD + 43° 4305	22 44 .7	+ 44 5	0.197	0.83	237	− 2	10 .05	dM 5 e
41⁵)	AC + 79° 3888	11 44 .6	+ 78 58	0.196	0.87	57	− 119	10 .9	sdM 4

¹) Spectroscopic binary.
²) Near companion measured by gravitation.
³) Probably unseen near companion.
⁴) White dwarf.
⁵) Subdwarf.
⁶) Flare star.
Nr. 7 B = UV Cet; Nr. 23 B = DO Cep; Nr. 30 = AD Leo; Nr. 40 = EV Lac.

B25. THE MOTION OF THE HYADES (L)

The Problem

This celebrated cluster near Aldebaran presents a special interest, because its stars are spread over an exceptionally big area of the sky, while their proper motions are considerable and have been measured very accurately. Moreover the radial velocities are well determined.

Because the stars of the cluster travel through space along parallel paths, it is possible to find the direction of this motion. This being determined, we shall derive the accurate distance of the cluster.

Procedure

1. Figure 61 shows a graph of δ against $\alpha \cos \delta$ for the main part of the cluster. The proper motions with components $\mu_\alpha \cos \delta$ and μ_δ are represented by arrows. Towards what point do these vectors converge?

One would wish to have the arrows drawn on a bigger scale. But this would not help much, since the spread, due to errors of measurement, is appreciable. Select stars of which the proper motion looks well representative for the surrounding group, and produce the vector towards the left side.

A number of such lines will meet roughly inside a limited area of which you estimate the centre of gravity. Read the coordinates of this *radiant*.

In precise research, one has to take into account that actually we have to produce arcs on a sphere, and not straight lines in a plane. The meeting point is determined by least squares.

Look up the literature, and compare your values with those, derived by least squares. For all the following calculations use the 'official' position.

2. In the table below you find the essential data for a few cluster stars, of which the results showed only very small probable errors. Select one of them. Its angular distance ϑ to the radiant might be read approximately from the diagram. However, working more carefully, we have to take into account that we are not working on a plane, but on the surface of a sphere; the distance is then quickly computed from the fundamental formula of spherical trigonometry (A3). In order to facilitate the further operations, this distance ϑ is already given in the table.

3. For a few stars, read in our table the radial velocity and compute the space velocity $V = V_{rad}/\cos \vartheta$.

The results are remarkably similar for all stars of the cluster; discussion shows that the remaining differences are mainly due to measuring errors. A mean value of 44.0 km/sec is now accepted.

4. For the star which you have selected compute the tangential velocity $V \sin \vartheta$.

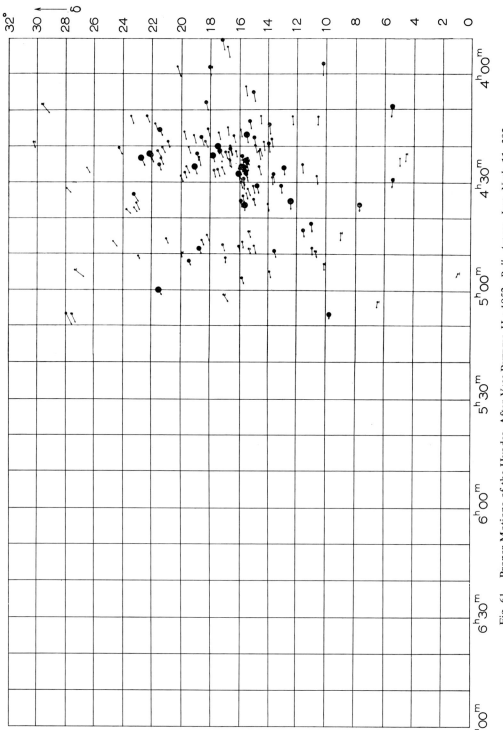

Fig. 61. Proper Motions of the Hyades. After VAN BUEREN, H.: 1952, *Bull. Astron. Inst. Neth.* **11**, 392.

TABLE

Some data about Hyades stars

Nr.	α	δ	m_v	$\mu_\alpha \cos\delta$	μ_δ	μ	ϑ	$V_{\rm rad}$
				(unit $0''.001/a$)				km/sec
6	3h 50m 15s	17° 10′ 47″	5.96	149 ± 2	− 28 ± 2	151	37°.5	31.6 ± 0.6
14	4 8 40	5 23 40	5.71	152 ± 2	+ 10 ± 2	152	32.5	35.8 ± 2.6
33	4 17 46	14 58 38	5.27	112 ± 2	− 23 ± 2	114	30.5	36.1 ± 0.8
74	4 25 55	17 10 35	8.2	106 ± 5	− 46 ± 6	115	29.0	40.5 ± 1.4
104	4 35 22	12 24 44	4.30	103 ± 3	− 11 ± 2	103	26.0	44.4 ± 2.1
112	4 43 15	11 36 57	5.43	74 ± 4	− 4 ± 3	74	24.0	38.2 ± 1.4
129	5 0 6	21 31 13	4.70	68 ± 1	− 42 ± 1	79	23.5	42.5 ± 1.4
131	5 6 37	27 58 7	6.1	62 ± 4	− 70 ± 3	94	26.5	41.3 ± 0.9

5. For this same star the proper motion is found from $\mu = \sqrt{\mu_\alpha^2 \cos^2 \delta + \mu_\delta^2}$ seconds of arc a year. Convert into a tangential velocity of 4.74 μ/p km/sec, where p is the distance in parsec.

6. By equating the results of (4) and (5) you find:

$$p = 4.74 \, \mu/(V \sin \vartheta)$$

(7.) Repeating this for several cluster stars, not much different results are found, from which a mean distance of the cluster is derived. Compute also the mean distance modulus. However, the results are of a sufficient precision to reveal individual distance differences and to yield an estimate of the size of the cluster in the line of sight. Among the stars, most distant from the centre, we mention Nr. 14 and Nr. 112.

Compute also their distance. Compare the extent of the cluster in depth to the extent across the line of sight.

References

SMART, W. M.: 1965, *Stellar Kinematics*, London, chapter 7.
VAN BUEREN, H.: 1952, *Bull. Astron. Inst. Neth.* **11**, 385.
WOOLLEY, R.: 1965, in *Stars and Stellar Systems*, Vol. V, p. 95.

Preparation

For each student: slide rule; ruler; trigonometric tables.

B26. THE MEAN LIGHT CURVE OF A CEPHEID (L)

The Material

A series of classical photo-electric observations of Cepheids have been made by J. Stebbins and his collaborators. By the interposition of suitable filters the brightness variations were observed in 6 different colours.

We study the measurements on η Aquilae, one of the first variable stars discovered and the prototype of cepheid stars with 'humps' in their light curves. The reality of such humps was definitely proved by photo-electrical measurements. This requires some care in the establishment of the precise light curve, but makes the problem especially interesting. We select the ultraviolet measurements corresponding to a mean wavelength of 3530 Å, because in this spectral range the amplitude of the variations is greatest.

The table below gives the complete series of observations. Data in parentheses were considered to be aberrant and were rejected. – The days are counted in the Julian era, introduced by J. J. Scaliger in 1582. It is a continuous numeration of the successive days, starting on noon of the first of January, 4712 B.C. For work on variable stars this dating is very practical and generally used. The *Astronomical Ephemeris* indicates how the Julian date (J.D.) corresponds to that of our civil calendar.

The Problem

1. Let us first use the main bulk of the measurements obtained between J.D. 3070.947 and 3225.704. Plot them on a suitable scale; the oscillations should not be pictured too sharp, but the whole of the data must be plotted. Keep to the tradition and let the *m*-scale increase downward. It is seen that the brightness fluctuates by more than 1^m. Apparently the observations include many cycles, but the curve is incomplete and frequently interrupted, due to cloudy weather, to the hours of the day, etc. However, since the variability of such stars is known to be almost perfectly periodical, we shall try to combine all observations into a mean curve, describing one cycle. This is a kind of puzzle, which will be solved step by step.

Procedure

2. At the very beginning you find some ascending branches, from which you derive a first rough approximation p_0 for the period. Then find a somewhat farther-off ascending branch; the distance between two ascending branches being a whole number of periods, you are able to obtain a more precise value p_1.

3. From all dates of the 2nd cycle subtract p_1; from the dates of the 3rd cycle

TABLE

Photoelectric Measurements of η Aql

JD 2430000 +	U 3530 Å
1998.947	$+1^m.07$
1999.945	$+0$ .28
2000.946	-0 .42
2001.936	-0 .11
2031.908	$+0$ .16
2032.912	$+0$ .60
2054.799	$+0$ .74
2055.784	$+1$ .00
3070.947	-0 .20
3074.916	$+1$ .05
3076.914	-0 .05
3082.841	$+0$ .96
3083.894	$+0$ .09
3084.880	-0 .34
3090.869	$+0$ .24
3098.859	-0 .36
3104.876	$+0$ .66
3117.783	$+1$ .02
3123.741	$+0$ .63
.871	$+0$ .64
3131.791	$+0$ .88
3140.829	$+0$ .52
3144.859	$+0$ .40
3151.823	$+0$ .26
3158.822	$+0$ .22
3172.801	$+0$ .22
3177.709	-0 .34
.804	-0 .39
.842	-0 .41
3179.685	$+0$ .16
3192.690	-0 .31
.756	-0 .30
3193.679	$+0$ .09
.719	$+0$ .10
3200.689	$+0$ .03
.722	$+0$ .02
3206.684	-0 .39
.718	-0 .38
3207.682	-0 .07
3217.644	$+0$ .78
3221.644	-0 .24
3225.640	$+1$ .04
.704	$(+1$ .02)

subtract $2p_1$, etc. All observations are now provisionally reduced to one cycle.

4. Plot all reduced points on a large scale. If p_1 has been over- or underestimated, the curves corresponding to the successive cycles are found to shift gradually back-

ward or forward. It may be found practical to distinguish between the several curves by using dots, or crosses, or other characteristic signs. If a systematic shift is observed, it will go on increasing.

Make again an estimate p_2 of the period; two decimals should now be well ascertained.

5. By comparing early measurements J.D. 1998.947–2055.784 with some of the later measurements, the period is found with a precision of $0^d.001$ or better.

6. Finally reduce again all observations with the definitive value p of the period, plot all reduced measurements and draw the mean light curve. Discuss critically the reality of the hump. Compare with the family of light curves, given by Hertzsprung.

(7.) Determine, as precisely as possible, the moment of the maximum. For this purpose draw a few horizontal chords, intersecting the light curve in two points each. Take on each chord the middle M of these points and connect $M_1, M_2, \ldots$ by a smooth curve. Its intersection with the light curve yields the maximum position.

TABULATION

	n	np_1	np
$p_0 = \ldots\ldots$	1	$\ldots\ldots$	$\ldots\ldots$
$p_1 = \ldots\ldots$	2	$\ldots\ldots$	$\ldots\ldots$
$p_2 = \ldots\ldots$	3	$\ldots\ldots$	$\ldots\ldots$
$p = \ldots\ldots$	4	$\ldots\ldots$	$\ldots\ldots$

J.D.	m	J.D. $- np_1$	J.D. $- np$
$\ldots\ldots$	$\ldots\ldots$		
$\ldots\ldots$	$\ldots\ldots$		
$\ldots\ldots$	$\ldots\ldots$	$\ldots\ldots$	$\ldots\ldots$
$\ldots\ldots$	$\ldots\ldots$	$\ldots\ldots$	$\ldots\ldots$

References

HERTZSPRUNG, E.: 1926, *Bull. Astron. Inst. Neth.* **3**, 115.
STEBBINS, J., KRON, G. E., and SMITH, J. L.: 1952, *Astrophys. J.* **115**, 292.

Preparation

For each pair: a few sheets of square coordinate paper.

In order to avoid the tedious computation of the products np_1, np_2, np a hand-computing machine may be used. Or else, take J. Peters: 1909, *Nouvelles Tables de Calcul*, Berlin.

B27. THE OBSERVATION OF DOUBLE STARS (S)

Among the numerous double stars there are a certain number of binaries, the components of which are close to each other in space and revolve about their common centre of gravity; the wider pairs, however, are for the greater part composed of stars which are at huge distances from each other, but from the earth are seen accidentally in line with each other. – The observation of double stars may lead to the determination of orbits. It may also be of interest in order to test an astronomical telescope.

For this last purpose we select by preference pairs for which the components do not differ by more than 2^m or 3^m. A list of suitable double stars is given below. Those which are most easily found with respect to their surroundings have been marked by an asterisk; some interesting cases of colour contrast between the components are marked by the note *col*.

Procedure

1. First note the number of your telescope, the diameter of its objective, and the ocular used. Direct the telescope to any bright star and focus.

2. Look to see which parts of the sky and which constellations are easily observable for the moment and are at a convenient height above the horizon. Select in the list of double stars a suitable wide pair, study the surrounding field in the *Star Atlas* and make a sketch. If the telescope has graduated circles, they may help in finding the object.

3. Direct your telescope to the double star and decide whether the components are *clearly separated, hardly separated* or *not separated*. Notice also whether the two components have a different colour: this will be found to correspond with the difference in spectral type, enhanced by contrast effects.

Make sure that you have identified the star correctly!

Note the name of the star, the distance of the components, m_1 and m_2.

4. Repeat the observations with other pairs, till you are able to estimate the actual resolving power of your telescope.

(5.) Take also one or two pairs with a considerable magnitude difference.

6. Compare your results with the theoretical resolution: $d'' = 13/p$, where p is the objective diameter in centimetres. This formula is to be applied only in the case of almost equally bright components. More general is the empirical expression:

$$d'' = \frac{13}{p}(1 + 0.2\overline{\Delta m}^2),$$

where Δm is the magnitude difference of the components.

List of Wide Double Stars

Name		α			δ	m_1-m_2	d''
	ψ^1 Psc	01^h	03^m	$.0$	$+21°.2$	5.6–5.8	30″
	ζ Psc	01	11	.1	$+\ 7$ .3	5.6–6.5	24″
*	$\gamma^{1,\,2}$ Ari	01	50	.8	$+19$ .0	4.8–4.8	8″.4
*	γ And	02	00	.8	$+42$ .1	2.3–5.1	10″
*	59 And	02	07	.8	$+38$ .8	6.0–6.7	17″
32 Eri = w	Eri	03	51	.8	-03 .1	5.0–6.3	7″
*	ϑ Tau	04	25	.6	$+15$ .8	3.6–4.0	336″
*	σ Tau	04	36	.2	$+15$ .8	4.9–5.3	426″
*	λ Ori	05	32	.4	$+09$ .9	3.7–5.6	4″
*	δ Ori	05	29	.4	-00 .3	2.0–6.8	53″
*	ι Ori	05	33	.0	-05 .9	2.9–7.3	11″
*	σ Ori	05	36	.2	-02 .6	3.8–7.2	13″
	γ Lep	05	40	.3	-22 .5	3.8–6.4	95″
	41 Aur	06	07	.8	$+48$ .7	6.1–6.8	8″
	β Mon	06	26	.4	-07 .0	4.7–5.2	7″
	19 Lyn	07	18	.8	$+55$ .4	5.6–6.5	15″
	ζ Cnc	08	09	.3	$+17$ .8	5.1–6.0	6″
	ι Cnc	08	43	.7	$+29$ .0	4.2–6.6	31″ col.
*	6 Vul	09	26	.9	$+24$ .6	4.6–6.0	403″
*	γ Leo	10	17	.2	$+20$ .1	2.6–3.8	4″
*	54 Leo	10	52	.9	$+25$ .0	4.5–6.3	6″
	τ Leo	11	25	.5	$+03$ .6	5.4–7.0	90″
*	γ Vir	12	39	.1	-01 .2	3.6–3.6	6″
	α CVn	12	53	.7	$+38$ .6	2.9–5.4	20″ col.
	17 CVn	13	07	.8	$+38$ .8	5.0–6.0	290″
*	ζ UMa	13	21	.9	$+55$ .2	2.4–4.0	14″
	κ Boo	14	11	.7	$+52$ .0	4.6–6.6	13″
Σ 1835 = ADS 9247	Boo	14	20	.9	$+08$ .7	5.1–6.6	6″
	π Boo	14	38	.4	$+16$ .6	4.9–5.8	6″
	α Lib	14	47	.8	-15 .8	2.9–5.3	228″
Σ 1962 = ADS 9728	Lib	15	36	.0	-08 .6	6.5–6.6	12″
*	β Sco	16	02	.5	-19 .7	2.9–5.2	14″ col.
	κ Her	16	05	.8	$+17$ .2	5.3–6.5	29″
	σ CrB	16	12	.8	$+34$ .0	5.7–6.7	6″
	ν Dra	17	31	.2	$+55$ .2	5.0–5.0	62″
*	ψ Dra	17	42	.8	$+72$ .2	4.9–6.1	30″
	95 Her	17	59	.4	$+21$ .6	5.1–5.2	6″
	100 Her	18	05	.8	$+26$ .1	5.9–6.0	14″
*	ε Lyr	18	42	.7	$+39$ .6	4.8–4.4	210″
*	ζ Lyr	18	43	.0	$+37$ .5	4.3–5.9	44″
*	β Lyr	18	48	.2	$+33$ .3	3.4–6.7	46″
	ϑ Ser	18	53	.8	$+\ 4$ .1	4.5–5.4	22″
*	β Cyg	19	28	.7	$+27$ .9	3.2–5.4	35″ col.
	α Cap	20	15	.1	-12 .7	3.8–4.6	378″
*	γ Del	20	44	.3	$+16$ .0	4.5–5.5	10″
	61 Cyg	21	04	.7	$+38$ .5	5.5–6.3	25″
	μ Cyg	21	41	.9	$+28$ .5	4.7–6.1	216″
	ξ Cep	22	02	.3	$+64$ .4	4.6–6.5	7″

References

AITKEN, R. G.: *The Binary Stars*. Reprinted in 1964, Dover Publications, New York.
GROOSMULLER, J. H.: 1935, *Hemel en Dampkring* **33**, 210, 390, 425; 1937, **35**, 57.

Preparation

For each pair: experimental telescope; flashlight; *Star Atlas.*
  Other suitable pairs may be found in:
A. Bečvář: 1964, *Atlas Coeli II, Katalog* 1950.0, Praha, p. 131.
G. D. Roth: 1960, *Handbuch für Sternfreunde*, Berlin.
Colour contrasts are listed in Kulikowski: 1961, *Spravočnik*, Moscow, p. 406.

B28. THE ORBIT OF A VISUAL BINARY (L)

(Method of Zwiers)

Introduction

Our material is a reproduction of visual observations obtained by a number of observers in the 19th and 20th century. The brightest component F is always taken as the origin; for the fainter component B the observer has measured the distance to F and the direction of FB. We only consider the *relative orbit*.

Take a sheet of tracing paper and mark the observed points. Draw as well as possible an ellipse between them: this is the *apparent orbit*, which corresponds to the true orbit, deformed by skew projection on the plane of the sky. In general you will notice that the main star does not occupy the focus of the apparent orbit.

We intend to reconstruct the true orbit by a graphical method. The drawing must be made very carefully if some precision in the results is to be expected.

Procedure (Figure 62)

1. The centre C of the true orbit is also the centre of the apparent orbit. This we determine by drawing a few very thin parallel lines (in any direction), then by drawing the straight line which divides them into equal segments ('the conjugate diameter').

2. By drawing the line CFP, which is the *projected major axis*, we find the *periastron P*, where the two stars have their closest approach.

3. The *eccentricity* is $e = (CF)/(CP)$, for this ratio is invariant for a skew projection.

4. The minor axis is the diameter, conjugate to the major axis. This remains true in projection; you are thus able to construct the *projected minor axis*.

5. *The auxiliary circle and the auxiliary ellipse.* The true, not projected orbit, with its (as yet unknown) axes a and b, may be obtained in principle by assuming an *auxiliary circle* of radius a, of which all y-coordinates are foreshortened in the proportion b/a. Inversely we may draw in our figure some chords parallel to the projection of the minor axis, and increase their lengths in the proportion $a/b = 1/\sqrt{1-e^2}$. Their extremities are connected by the *auxiliary ellipse*, dotted in Figure 62, which is the projection of the auxiliary circle. (Here inaccuracies in the data or in the drawing have a considerable influence on the results and on the subsequent derivations; careful and critical work is required.)

6. We must now determine the position of the auxiliary ellipse in space. It is clear that the major axis of this ellipse is the line around which the orbit plane has been tilted. Construct the two axes α, β of the auxiliary ellipse by drawing a circle around C which intersects this ellipse; by connecting the intersections you find the direction of α and β. Measure α, β and compute the *inclination i of the orbit plane* to the sky from $\beta/\alpha = \cos i$ (cf. (13)).

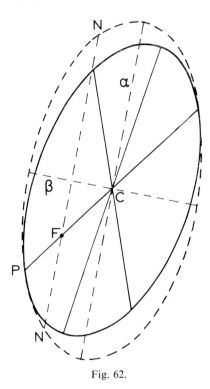

Fig. 62.

7. The only radius vector which remained unaffected by the projection is α; consequently this is equal to the *major axis a*. Reduce this to seconds of arc.

8. The *line of nodes NN* passes through *F* and must be parallel to α. Read the position angle Ω of the node; select that node for which $0 < \Omega < 180°$.

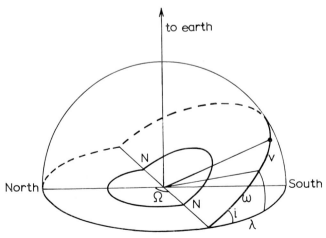

Fig. 63 (after L. Binnendijk).

9. The *longitude ω of the periastron P* is the angle *NFP* between the major axis and the line of nodes, measured in the direction of the motion (Figure 62); it is observed in projection as an angle λ (Figure 63). Imagine a sphere around *F* and take the plane of the sky as the plane of our drawing. Compute ω by applying the formula for a right-angled spherical triangle: $\tan \lambda = \tan \omega \cos i$.

10. Finally, determine approximately from the observations the *period of revolution T* and the moment of *periastron passage t*. Could they perhaps be found more precisely by interpolation?

11. Compare the elements determined by you with those ascertained by the observer himself or found in the literature.

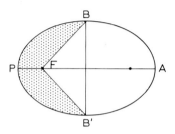

Fig. 64.

12. Check the period as follows (Figure 64). Call *B, B'* the extremities of the minor axis of the true orbit. Let T_1, T_2 be the time intervals needed to cover the sections *BPB'*, *B'AB* of the orbit. The hatched area is $PBB' - FBB' = \frac{1}{2}\pi ab - abe$. The remaining part of the ellipse area is $\frac{1}{2}\pi ab + abe$. From Kepler's law we have:

$$\frac{T_1}{T_2} = \frac{\frac{1}{2}\pi ab - abe}{\frac{1}{2}\pi ab + abe} = \frac{\pi/2 - e}{\pi/2 + e}.$$

If therefore T_1 or T_2 alone are known, it is still possible to find the period $T = T_1 + T_2$.

(13.) If the auxiliary ellipse has only a small eccentricity, measure the diameters in several directions at intervals of 15°. Through these points draw a sine line, determine the zero points, they are symmetrically located with respect to the maximum and the minimum.

(14.) Knowing the parallax, you may compute *a*, *b*, *FP*, *FA* in AU.

References

AITKEN, R. G.: 1935, *The Binary Stars*, New York. (Reprinted in 1964, Dover Publ.)
BINNENDIJK, L.: 1960, *Properties of Double Stars*, Philadelphia, Chapter II.
McLEOD, N. W.: 1939, *Astron. J.* **48**, 119.
VAN DE KAMP, P.: 1958, in *Handbuch der Physik* **50**, pp. 187 and 196.

Preparation

For each student: a sheet of tracing paper.

For several suitable pairs, the older observations are found represented on a convenient scale, ready for use, in the monograph of T. J. J. See: 1896, *Researches on the Evolution of Stellar Systems*, vol. I, Lynn. – Much more recent is the work of A. Aa. Strand, but the diagrams should be magnified before they are handed to the students.

We recommend:

ζ Her $= \Sigma$ 2084 $=$ ADS 10157; data from See, 1955, 'Recent Measurements', *Astron. J.* **60**, p. 39. – Par. 0".105

ξ UMa $= \Sigma$ 1523 $=$ ADS 8119; data from See, 1955, 'Recent Measurements', *Astron. J.* **46**, p. 193. – Par. 0".124

η CrB $= \Sigma$ 1937 $=$ ADS 9617; data from See. – Par. 0".066

γ Vir $= \Sigma$ 1670 $=$ ADS 8630; data from See and from Strand: 1937, *Ann. Obs. Leiden*, **18**^{2}, p. 89. – Par. 0".103

70 Oph $= \Sigma$ 2272 $=$ ADS 11046; data from See and from Strand: 1937, *Ann. Obs. Leiden*, **18**^{2}, p. 137. – Par. 0".196

ζ Boo $= \Sigma$ 1888 $=$ ADS 9413; data from See and mainly from Strand: 1947, *Ann. Obs. Leiden*, **18**^{2}, p. 97. – Par. 0".029

Numerical data can be plotted on rectangular or on polar coordinate paper, depending on the coordinates published.

As a demonstration, the student should be given the opportunity to measure the position angle and the separation for an easy pair by means of the micrometer, adapted to a telescope.

The Principle

Let μ_1 and μ_2 be the masses of the components of a binary, expressed in solar masses. From Newton's law:

$$\mu_1 + \mu_2 = \frac{a^3}{P^2 p^3},$$

where a is the angular semi-major axis of the relative orbit; $p =$ the parallax, in seconds of arc, $P =$ the period of revolution in years.

We would be able to compute the parallax if we knew the masses. These we may derive from the mass-luminosity law, provided the apparent magnitude is reduced to absolute magnitude via an approximate value for p. Obviously we shall have to apply successive approximations. Our equations are:

$$\rightarrow \mu_1 + \mu_2 = \frac{a^3}{P^2 p^3}$$
$$M = m + 5 + 5 \log p.$$
$$M = f(\mu).$$

Procedure

First study the binary of which you have drawn the orbit (B28). Then take one or two of the following pairs:

Star	Visual		P	a	Spectrum		Trig. par.
	m_1	m_2					
α Cen	−0.04	1.38	79.9	17".58	G2	K5	0".751
η Cas	3.47	7.22	480.0	11".99	G0	M0	0".176
ε Hyd	3.7	4.8	15.0	0.21	G0	?	0".014

1. First assume that the components have a mass equal to that of the sun, and write $\mu_1 = \mu_2 = 1$.
   From the law of gravitation, calculate a provisional value of p.
2. Determine provisionally the absolute magnitudes from the relation

$$M = m + 5 + 5 \log p.$$

3. Reduce these visual magnitudes to bolometric ones, by applying the bolometric corrections m listed at the end of B21. Then from the mass-luminosity law determine better approximations for μ_1 and μ_2. Or use directly the relation between mass and absolute visual magnitude (see the table below).

4. Repeat the operations (1), (2), (3) with these better values. Go on till the values obtained do not change appreciably.

5. Compare your results with the parallax, as directly determined by trigonometrical measurements.

(6.) It is interesting to see how quickly the method converges. Take as a starting point $\mu_1 = \mu_2 = 10$ and watch the successive approximations.

TABLE

The Mass-Luminosity Law
(C. W. ALLEN: 1963, *Astrophysical Quantities*, p. 203)

$\log \mu$	M_{bol}	M_v
-1.0	$+11.3$	$+14.8$
-0.8	$+10.3$	$+13.7$
-0.6	$+\ 9.4$	$+12.4$
-0.4	$+\ 8.1$	$+10.6$
-0.2	$+\ 6.6$	$+\ 7.8$
0.0	$+\ 4.7$	$+\ 4.8$
$+0.2$	$+\ 2.7$	$+\ 2.8$
$+0.4$	$+\ 0.8$	$+\ 1.2$
$+0.6$	$-\ 0.9$	$-\ 0.1$
$+0.8$	$-\ 2.4$	$-\ 1.2$
$+1.0$	$-\ 3.9$	$-\ 2.5$
$+1.2$	$-\ 5.4$	$-\ 3.7$
$+1.4$	$-\ 6.8$	$-\ 4.8$
$+1.6$	$-\ 8.1$	$-\ 5.9$
$+1.8$	$-\ 9.5$	$-\ 7.0$

Reference

A monograph on this subject is: H. N. Russell and Ch. E. Moore: 1940, *The Masses of the Stars*; it explains how the method has been extended to cases where only part of the orbit has been observed (p. 3) and contains a catalogue (p. 180 and 183).

Preparation

More examples may be chosen from Ch. E. Worley: 1963, *Public. U.S. Naval Obs.* **18,** part 3.

B30. THE ORBIT OF A SPECTROSCOPIC BINARY (L)

(2nd Method of Henroteau)

The Problem

Given: a set of measurements of the radial velocity, referring to one component of a spectroscopic binary with respect to the other one. The original data have already been reduced to normal places and corrected for the relative motion of the earth and the sun. To derive the orbit in space.

Procedure

1. Plot the radial velocity as a function of time (*velocity curve*) $v_{rad}(t)$. In doing so, take into account the weight of the measurements, make the size of the dots roughly proportional to their weight. Choose a suitable scale for the coordinates! The curve must be smooth; the end of the period must fit fluently into the beginning of the next period. In the following we shall work as if i was 90°, which means that the direction of observation would be in the orbital plane. Actually the value of A which we shall obtain applies to $A \sin i$.

The *period T* is directly found by looking at the tabulated radial velocities.

2. Under the velocity curve we draw the *integral curve*:

$$z = \int_0^t v_{rad} \, dt,$$

which measures the position of the star in the line of sight as a function of time. To obtain this, determine the increase of the area with intervals of 2 days; simply add successively the ordinates, corresponding to the mean moment of each interval

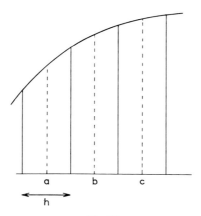

Fig. 65.

(Figure 65):
$$h(a + b + c + \cdots).$$

Where the curvature is strong, the intervals should be diminished to one day or even less. – (If you have not made a graphical integration earlier, start with a part of the curve where the ordinates are positive!)

If the points O and T of the integral curve are not at the same height, thus if:

$$\int_O^T v_{\text{rad}}\, dt \neq 0,$$

this would show that the radial velocity of the system as a whole has not been completely eliminated. In that case connect the points corresponding to the moments O and T, and reduce this line to the horizontal by applying a small correction to all points of the integral curve (Figure 66).

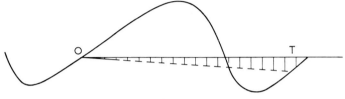

Fig. 66.

3. Let us imagine the orbit (as yet unknown), which is an ellipse symmetric with respect to its centre C. The integral curve measures the distance z along the direction of observation, reached by the secondary star in the orbital plane.

In the course of our argument we consider at the same time *the radial velocity curve*, the *integral curve*, the *orbit*, and the *line of sight coordinate z* (Figure 67).

The centre of the ellipse corresponds to a point C' or C'', halfway between the maximum and the minimum of the integral curve.

4. From Kepler's laws it may be shown that the radial velocity is greatest in the

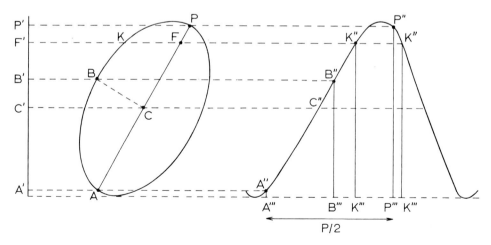

Fig. 67.

nodes. The maxima of the radial velocity curve thus correspond to the moments when the star reaches the *nodes K*. At these same moments K'' in the integral curve the ordinates K''' K'' should obviously be equally high; if this is not quite the case, this is due to the uncertainty in the moments of maximum and minimum velocity. Select the best mean value. The moments when the distance of the nodes has been reached are also the moments when the distance of the *focus F* is attained.

5. *Apastron A* and *periastron P* satisfy two conditions: (a) their abscissae are at a distance in time equal to $T/2$; (b) their coordinates z, measured from the line CC'', are equal but of opposite sign.

Find by trial and error which points satisfy these conditions. In between them there should always be a maximum or a minimum of the integral curve. It may help to know that the periastron is always at the same side of the line CC' as F and it is located on that side of the integral curve where the slope is the steeper one.

6. Determine the *eccentricity*

$$e = \frac{CF}{CP} = \frac{C'F'}{C'P'}.$$

7. We now determine the extremity B of the minor axis, by calculating the time interval $t - t_0$ which the star needs for the arc PB of the orbit.

In B the eccentric anomaly is $E = \pi/2$. From Kepler's equation:

$$M = \frac{\pi}{2} - e = \frac{2\pi}{T}(t - t_0).$$

or

$$t - t_0 = \frac{T}{2} \times \left(\frac{1}{2} - \frac{e}{\pi}\right).$$

We thus know the distance $P''' B'''$ and the points B''', B'', B'.

8. We have still to find ω, the *longitude of the periastron* (PFK or KFC).

$$P'C' = PC \cdot \sin \omega$$
$$B'C' = BC \cdot \cos \omega$$
$$\operatorname{cotan} \omega = \frac{B'C'}{P'C'} \cdot \frac{PC}{BC} = \frac{B'C'}{P'C'} \frac{1}{\sqrt{1 - e^2}}.$$

9. Finally we find the major axis $a = PC = P'C'/\sin \omega$.

10. Calibrate the z scale by realizing that the elements of area in the radial velocity curve represent elements of velocity $\times$ time $=$ distance in the z direction. Derive the value of the major axis a. Actually, this is rather the value of $a \sin i$; the angle i cannot be determined by spectroscopic measurements alone.

(11.) Draw the elliptical orbit thus determined (putting $i = 90°$).

References

AITKEN, R. G.: 1935, *The Binary Stars*, New York (Reprinted in 1963, Dover Publ.), chapter VI.
BINNENDIJK, L.: 1960, *Properties of Double Stars*, Philadelphia, Chapter IV.
HENROTEAU, F. C.: 1928, in *Handbuch der Astrophysik* **6**, p. 369.
STRUVE, O. and HUANG, S. S.: 1958, in *Handbuch der Physik*, Vol. 50. Springer, p. 243.

TABLE

Data for Two Spectroscopic Binaries

Star: HR 6385	Star: HR 8800
1920, *Publ. Domin. Astroph. Obs.* **1**, 199	1920, *Publ. Domin. Astroph. Obs.* **1**, 241
Corrected for system velocity	Corrected for system velocity
$\gamma = +3.51$ km/sec	$\gamma = 15.13$ km/sec

Phase (d)	v_{rad}	Weight	Phase (d)	v_{rad}	Weight
0.26	− 26.0	1.5	0.203	− 99.7	1
0.94	− 38.3	1.5	0.527	− 113.5	$\frac{1}{4}$
1.42	− 34.5	2.0	0.570	− 76.8	$1\frac{1}{2}$
2.56	− 29.2	2.5	0.976	− 42.8	$\frac{1}{2}$
4.42	− 19.5	3.0	0.987	− 34.4	1
6.46	− 6.9	3.5	1.114	− 24.8	$\frac{1}{2}$
7.80	− 0.9	3.0	1.331	+ 7.1	1
10.12	+ 5.2	2.5	1.584	+ 46.4	$\frac{1}{4}$
13.46	+ 14.9	3.5	1.747	+ 40.6	$\frac{1}{2}$
16.02	+ 18.5	5.0	1.986	+ 64.3	$\frac{1}{2}$
17.49	+ 19.1	3.5	2.016	+ 63.8	1
19.69	+ 19.6	3.5	2.590	+ 64.1	$\frac{1}{2}$
21.34	+ 4.0	2.0	2.681	+ 70.0	$1\frac{1}{2}$
21.84	− 0.3	1.0	2.976	+ 22.3	$1\frac{1}{2}$
22.43	− 10.7	3.5	3.234	− 32.3	$\frac{3}{4}$
23.50	− 26.0	1.5	3.325	− 70.5	$\frac{1}{2}$
24.18	− 38.3	1.5	3.540	− 99.7	1
24.66	− 34.5	2.0	3.864	− 113.5	$\frac{1}{4}$
25.80	− 29.2	2.5	3.907	− 76.8	$1\frac{1}{2}$
27.66	− 19.5	3.0	4.313	− 42.8	$\frac{1}{2}$
29.70	− 6.9	3.5	4.324	− 34.4	1
31.04	− 0.9	3.0	4.451	− 24.8	$\frac{1}{2}$
33.36	+ 5.2	2.5	4.668	+ 7.1	1
36.70	+ 14.9	3.5	4.921	+ 46.4	$\frac{1}{4}$
39.26	+ 18.5	5.0	5.084	+ 40.6	$\frac{1}{2}$
40.73	+ 19.1	3.5	5.323	+ 64.3	$\frac{1}{2}$
42.93	+ 19.6	3.5	5.353	+ 63.8	1
			5.927	+ 64.1	$\frac{1}{2}$
			6.018	+ 70.0	$1\frac{1}{2}$

When plotting the curve, take:
2 days .... 1 cm
10 km/sec .... 2 cm

When plotting the curve, take:
1 day .... 4 cm
10 km/sec .... 0.5 cm

Preparation

The classical, widely followed method is that of Lehmann-Filhés. However, for this exercise we have preferred (with apologies!) a method of Henroteau, because it requires only a minimum of previous knowledge and is so easy to visualize.

Rich material is found in the *Publications of the Dominion Astrophysical Observatory*. We avoid those cases where the velocity curves are too symmetric and less interesting. We quote the volumes and pages where nice examples are to be found:

1, 153, 233, 281; **2**, 167, 263; **3**, 145, 151 (ζ Aur!), 204, 335, 349; **4**, 39, 85, 97, 108, 183; **6**, 88; **7**, 105, 415; **11**, 241.

Photographic reproductions of stellar spectra, showing the varying velocity shift beautifully, may be bought from the Yerkes Observatory; in the catalogue of reproductions they have the designation SS 171–179 and SS 191.

B31. THE ORBIT OF AN ECLIPSING VARIABLE (L)

The Problem

The light curve of an eclipsing variable yields a wealth of information concerning the geometrical and physical properties of the two component stars. We take, as an example, the fine measurements on W Delphini obtained by Wendell with a polarizing photometer. The 500 observations have been combined into 'normal points', of which we select about 25 (Table A). For the analysis we apply H. N. Russell's classical method. We assume for simplification that each of the stellar discs is uniformly bright.

TABLE A

The light curve of W Delphini (normal points)
After H. N. RUSSELL: 1912, *Astrophys. J.* **36**, p. 134.

$t - t_0$	m	$t - t_0$	m
$- 0^d.2894$	9.41	$+ 0^d.0356$	12.02
.2306	9.59	.0460	11.87
.1911	9.88	.0659	11.58
.1506	10.23	.1036	10.88
.1121	10.78	.1445	10.31
.0715	11.51	.1847	9.90
.0313	12.05	.2242	9.63
.0169	12.08	.2708	9.48
$- .0082$	12.07	.94	9.42
$+ .0060$	12.16	1.90	9.35
.0139	12.09	2.04	9.41
.0269	12.03	2.67	9.38

Procedure

1. Plot the observed magnitudes against time. As is usual in such work, the sense of increasing magnitude (and decreasing brightness) is taken downward. Draw carefully a smooth curve through the normal points. – A secondary minimum was not observed. The curve looks almost perfectly symmetric.

2. We notice directly that the brightness remains practically constant for more than one hour during the primary minimum. This means that the eclipse is either total or annular. We shall find later that it is total: the small component disappears behind the big one.

The period has been found to be $P = 4^d.8061$. For such narrow pairs the orbit may be considered to be circular (Figure 68). The time $t - t_0$, elapsed after mid-totality, corresponds to an angle $\vartheta = 360° \left[(t - t_0)/P \right]$. We transform the abscissae into this angle (4 decimals); in order to save time this has already been done in Table B. For a dozen

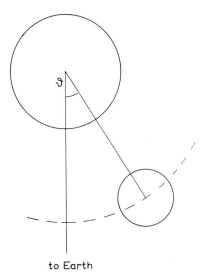

to Earth

Fig. 68.

values of m you find there the corresponding values of $t_1 = t - t_0$ as read from the descending branch of the curve, t_2 from the ascending branch.

The corresponding values of ϑ are practically equal and means have been taken. Check a few points roughly.

3. We shall also transform the ordinates (Figure 69). The maximum luminosity (= the total luminosity of the two uneclipsed components) we shall put equal to 1, the minimum equal to λ, the luminosity at a moment t equal to l. The loss of light for the star which is behind is $1 - \lambda$ at mid-totality, $1 - l$ at a moment of the partial

TABLE B
Analysis of the light curve of W Delphini

α	$1-l$	mag.	t_1	t_2	ϑ	$\sin^2 \vartheta$	$\sin^2 \vartheta - A$	$\psi(k, \alpha)$	k
0.0	0.0000	$9^m.400$	$-0^d.304$:	$+0^d.300$:	0.394:				
.1	.0917	9 .505	.2540	.2515	.3304				
.2	.1834	9 .620	.2285	.2258	.2968				
.3	.2750	9 .749	.2075	.2030	.2681				
.4	.3667	9 .896	.1884	.1830	.2426				
.5	.4584	10 .066	.1682	.1644	.2173				
.6	.5500	10 .266	.1486	.1470	.1931				
.7	.6417	10 .514	.1270	.1274	.1661				
.8	.7334	10 .835	.1070	.1048	.1381				
.9	.8250	11 .292	.0824	.0788	.1054				
.95	.8709	11 .624	.0655	.0624	.0836				
.98	.8985	11 .884	.0505	.0462	.0632				
.99	.9076	11 .985	.0430	.0390	.0536				
1.00	.9168	12 .100	$-$ .021:	$+$ .019:	.026:				

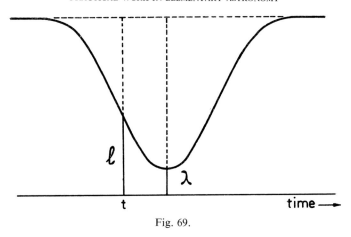

Fig. 69.

phase. The *fractional loss of light* $\alpha = (1-l)/(1-\lambda)$ is at the same time the occulted fraction of the star which is behind, expressed in its surface area. In order to facilitate this computation, there are tables by which a magnitude interval Δm is converted into $1 - l$. In Table B you find a column with the values of α; you understand that the magnitudes m in the third column have been selected in such a way that simple round values of α are found.

Check a few points of $1 - l$ and α.

4. The shape of the light curve depends on the parameters: $i =$ the inclination of the orbit plane on the celestial sphere, always close to $90°$; r_1 and r_2, the radii of the big star and of the small star; we may also say: r_1 and $r_2/r_1 = k$, where $k < 1$.

If these three parameters were known, it would be a rather simple geometrical problem to calculate the light curve $\alpha(\vartheta)$. Reversely, calculating a set of light curves for different values of the parameters, we could compare them to the observed curve and find which of them coincides. – But it is clear that the number of possible curves would be considerable. We must get rid of two out of the three parameters.

Given the three parameters, the luminosity of the system depends on Δ, the projected distance between the centres C_1C_2 of the stars. From the geometry of the problem (Figure 70), we see that $\cos D = \cos \vartheta \cdot \sin i$; consequently

$$\Delta^2 = \sin^2 D = 1 - \cos^2 \vartheta \cdot \sin^2 i = \cos^2 i + \sin^2 i \cdot \sin^2 \vartheta \qquad (1)$$

We may write: $\Delta/r_1 = \varphi(k, \alpha)$. For the functions α, k, Δ/r_1 are *ratios, not depending on the absolute size of the system*; thus:

$$r_1^2 \{\varphi(k_1, \alpha)\}^2 = \cos^2 i + \sin^2 i \sin^2 \vartheta.$$

Apply this equation to two fixed points of the light curve, for which we choose $\alpha_2 = 0.6$ and $\alpha_3 = 0.9$; and then again to a variable point α. We get:

$$r_1^2 \{\varphi(k, \alpha_2)\}^2 = \cos^2 i + \sin^2 i \times \sin^2 \vartheta_2 \qquad (a)$$

$$r_1^2 \{\varphi(k, \alpha_3)\}^2 = \cos^2 i + \sin^2 i \times \sin^2 \vartheta_3 \qquad (b)$$

$$r_1^2 \{\varphi(k, \alpha)\}^2 = \cos^2 i + \sin^2 i \times \sin^2 \vartheta \qquad (c)$$

By calculating

$$\frac{(a) - (b)}{(c) - (b)} = \frac{\{\varphi(k, \alpha)\}^2 - \{\varphi(k, \alpha_3)\}^2}{\{\varphi(k, \alpha_2)\}^2 - \{\varphi(k, \alpha_3)\}^2} = \psi(k, \alpha) =$$

$$= \frac{\sin^2 \vartheta - \sin^2 \vartheta_3}{\sin^2 \vartheta_2 - \sin^2 \vartheta_3} = \frac{\sin^2 \vartheta - A}{B} \qquad (2)$$

we have eliminated the variables r_1 and i. Only the parameter k is left.

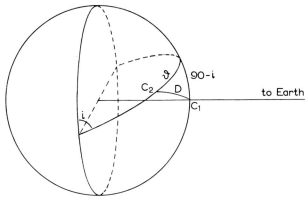

Fig. 70.

So, instead of studying the relation between α and ϑ, we shall look to the relation between α and $\psi = (\sin^2\vartheta - \sin^2\vartheta_3)/(\sin^2\vartheta_2 - \sin^2\vartheta_3)$, which involves only the one *parameter k*.

For our purpose it is sufficient to introduce the new variable ψ for a few 'sensitive' points: $\alpha = 0.20, 0.40, 0.80, 0.95$. The computation is easy; ϑ_2 and ϑ_3 are fixed numbers (corresponding to $\alpha = 0.60$ and $\alpha = 0.90$), and so are A and B. In this particular case $A = 0.0369$ and $B = 0.0258$.

TABLE C

The function $\psi(k, \alpha)$

k	$\alpha = 0.00$	0.20	0.40	0.60	0.80	0.90	0.95
1.00	9.350	3.852	1.455	0.000	− 0.805	− 1.000	− 1.048
0.95	8.309	3.474	1.333	0.000	− 0.783	− 1.000	− 1.068
0.90	7.470	3.170	1.236	0.000	− 0.764	− 1.000	− 1.085
0.85	6.779	2.921	1.155	0.000	− 0.749	− 1.000	− 1.098
0.80	6.196	2.709	1.087	0.000	− 0.736	− 1.000	− 1.111
0.75	5.696	2.528	1.028	0.000	− 0.725	− 1.000	− 1.122
0.70	5.262	2.371	0.977	0.000	− 0.715	− 1.000	− 1.132
0.65	4.881	2.232	0.932	0.000	− 0.706	− 1.000	− 1.141
0.60	4.542	2.109	0.892	0.000	− 0.698	− 1.000	− 1.150
0.55	4.239	1.998	0.856	0.000	− 0.690	− 1.000	− 1.158
0.50	3.965	1.898	0.823	0.000	− 0.683	− 1.000	− 1.165
0.45	3.716	1.806	0.792	0.000	− 0.676	− 1.000	− 1.172
0.40	3.489	1.722	0.764	0.000	− 0.670	− 1.000	− 1.179
0.35	3.281	1.644	0.738	0.000	− 0.664	− 1.000	− 1.185

Now let us compare the results! In table C you find the function $\psi(\alpha)$ for a number of values of k. For each of your five values of α you may determine, by interpolation, the best-fitting value of k. All these values should be identical. They are not, because the results of the analysis are very sensitive to observational errors (and because the assumption of uniform brightness is not correct). Take a mean value for k.

5. For the determination of the other elements of the orbit we have first to find the precise beginning of the eclipse and the precise beginning of totality (1st and 2nd contact). These are the moments when $\alpha = 0$ or $\alpha = 1$; while $\Delta = r_1 + r_2$ or $\Delta = r_1 - r_2$. Applying equation (1):

$$\left.\begin{array}{l} (r_1 + r_2)^2 = \cos^2 i + \sin^2 i \times \sin^2 \vartheta_1 \\ (r_1 - r_2)^2 = \cos^2 i + \sin^2 i \times \sin^2 \vartheta_2 \, . \end{array}\right\} \tag{3}$$

Equation (2) yields:

$$\sin^2 \vartheta = A + B\psi \, .$$

Our equations (3) become:

$$r_1^2 (1 + k)^2 = \cos^2 i + A \sin^2 i + B \sin^2 i \times \psi(k, 0)$$
$$r_1^2 (1 - k)^2 = \cos^2 i + A \sin^2 i + B \sin^2 i \times \psi(k, 1) \, .$$

We want to know i and r_1. Divide the terms of the first equation by those of the second one, so that r_1 is cancelled. Solve for $\sin^2 i$. Then find r_1.

It remains unknown towards which direction the plane is inclined over i with respect to the celestial sphere.

6. Finally draw the system to scale at the moment of mid-eclipse.

7. We still have to decide whether the observed minimum occurred when the small component was behind or in front of the big one (total or annular eclipse). – Suppose that the small component was behind at the secondary minimum, which was not observed and of which the dip was certainly less than 0.03. What can you tell about the luminosity of the small star? When it was in front of the big star, what fraction of the light of this star did it intercept? What should have been the ratio $k = r_2/r_1$? Does this check with your determination? What conclusion do you arrive at?

(8.) Compare the luminosities, the radii, the surface brightnesses for both stars. The spectral types are: A0 and G5.

References

BINNENDIJK, L.: 1960, *Properties of Double Stars*, Philadelphia, Chapter VI.
RUSSELL, H. N.: 1912, *Astrophys. J.* **35**, 315 and **36**, 133.
SHAPLEY, H.: 1912, *Astrophys. J.* **36**, 269.
TSESSEVICH, V. R.: 1947, in *Variable Stars*, Moscow, Vol. 3, ch. 5. (In Russian.)

Preparation

Eclipsing variables with similar light curves are: S Cnc, SW Cyg, U Cep, all of them discussed in SHAPLEY's (1912) paper.– Tables of logarithms and trigonometric functions (4 decimals); square coordinate paper.

B32. THE OBSERVATION OF STAR CLUSTERS
AND NEBULAE (S)

For many of the following objects, especially the nebulae, moonlight is prohibitive. Each object to be observed is to be found first in the *Star Atlas*. Memorize the configuration of the neighbouring stars, if necessary make a rough sketch.

1. *Observe some of the following objects first with a field glass, afterwards with your telescope*

The Pleiades have already been studied (B16).	$3^h\ 44^m\ +24°.0$
1. Praesepe, in the constellation Cancer.	$8^h\ 37^m\ +20°.2$
2. The Hyades, near Aldebaran.	$4^h\ 17^m\ +15°.5$
3. The double cluster in Perseus (*h* and *χ*).	$2^h\ 17^m\ +56°.9$
4. Coma Berenices in Coma.	$12^h\ 40^m\ +24°.0$
5. M31, the spiral nebula in Andromeda.	$0^h\ 37^m\ +40°\ 19'$
6. M42, the diffuse nebula in Orion.	$5^h\ 30^m\ -\ 5°\ 27'$
7. M35 in Gemini.	$6^h\ 05^m\ +24°\ 21'$
8. Cluster in Monoceros, to the left of ε Mon.	$6^h\ 30^m\ +\ 4°\ 54'$
9. M67 in Cancer.	$8^h\ 48^m\ +12°\ \ 0'$

2. *Observe some of the following objects with your telescope*

10. Open cluster in Cassiopeia, halfway between γ and k.	$0^h\ 40^m\ +61°\ 31'$
11. M34 in Perseus.	$2^h\ 39^m\ +42°\ 32'$
12. M38 in Auriga.	$5^h\ 25^m\ +35°\ 48'$
13. M37 in Auriga.	$5^h\ 49^m\ +32°\ 33'$
14. M41 in Canis Major.	$6^h\ 44^m.9 - 20°\ 42'$
15. M3, a globular cluster in Canes Venatici.	$13^h\ 40^m\ +28°\ 38'$
16. M5, a globular cluster in Serpens.	$15^h\ 15^m.9 +\ 2°\ 16'$
17. M13, a globular cluster in Hercules.	$16^h\ 39^m.9 +36°\ 33'$
18. M6, an open cluster in Scorpius.	$17^h\ 36^m.7 -32°\ 10'$
19. M8, a diffuse nebula in Sagittarius ('Lagoon nebula')	$18^h\ \ 0^m.6 -24°\ 23'$
20. M22, a globular cluster in Sagittarius.	$18^h\ 33^m.3 -23°\ 57'$
21. M15, a globular cluster in Pegasus.	$21^h\ 27^m.6 +11°\ 57'$
22. M2, a globular cluster in Aquarius.	$21^h\ 30^m.9 -\ 1°\ \ 4'$

Always put the telescope first at the required declination, but clamp it gently so that it remains possible to adjust it.

If possible, observe some of these clusters in a telescope of medium power (especially a globular cluster).

3. Make a careful drawing of the Pleiades. Indicate the E–W direction. Make a note of the type of telescope which you have used.

Then compare with a map, indicate the brightest stars by their names.

LITERATURE CONCERNING INTERESTING OBJECTS

First of all: *Norton's Star Atlas.*
Most books give too many objects for our purpose. We need a selection of a few representative objects, *easy to find.*
AHNERT, P.: 1961, *Beobachtungsobjekte für Liebhaberastronomen,* Leipzig.
MCKREADY, K.: 1912, *A Beginner's Star Book,* London.
MCKREADY, K.: 1913, *Sternbuch für Anfänger,* Leipzig.
More extensive lists are found in many books for the amateur astronomer. We especially mention:
BAILEY, S. I.: 1908, *Harvard Annals,* **60,** No. 8, reproduced in PLASSMANN, J.: 1922, *Hevelius,* Berlin, p. 357.
BRANDT, R.: 1956, *Himmelswunder im Feldstecher,* Leipzig.
VEHRENBERG, H.: 1968, *Atlas of Deep Sky Splendors,* Cambridge, Mass.
Very useful finding charts are inserted in the books of MCKREADY and of BRANDT.

Maps of the Pleiades are found in: *Norton's Star Atlas,* map 5 (insert); Kulikowski: 1961, *Spravočnik* (3rd ed.), p. 406; L. Binnendijk: 1946, *Ann. Obs. Leiden* **19,** p. 119.

B33. OPEN CLUSTERS (L)

The Problem

We want to compare the colour-magnitude diagrams of several open clusters (Figure 71). These diagrams will of course differ because of the different distances. Apart from this effect, one might expect to find in all of them the same relation between colour and magnitude, as long as the stars have not evolved away from the 'zero age main sequence' (ZAMS) and changed their internal model. We shall have to look for this most interesting evolutionary effect!

Procedure

1. As a starting point, we take the colour-magnitude diagram of the *Hyades*. The stars were measured in the U-B-V system; reductions for interstellar reddening are unimportant. The distance of this moving cluster is very precisely known from the proper motions and radial velocities (B25); it has even been possible to determine individual distances for each star and to reduce their brightness individually to absolute magnitudes M_v. These data are more recent and slightly more precise than those used in B25.

Above the main ridge-line of the diagram, over a height not exceeding $0^m.75$, there is a sparsely populated fringe of dots, probably corresponding to binary stars, not recognized as such.

Mark carefully on tracing paper the ridge line of the diagram and also copy the scale of coordinates.

2. Next we consider the diagram for the *Pleiades*. For colour indices exceeding $+0.8$, there is some spread and not much weight should be given to the measurements. Filled circles represent stars whose membership in the cluster is most probable. Reddening and absorption are negligible.

Transfer your copy of the Hyades diagram to the Pleiades diagram; let the scale of colour indices coincide and try to obtain the best possible coincidence by a shift in the vertical direction, which will take account of the distance ratio. If deviations between both diagrams are unavoidable, let them occur at the left side, where the stars have a greater mass and might already have started an evolution away from the main sequence.

Trace the extension of the ridge line to the left, as far as this is possible. Note carefully the shift, expressed in magnitudes, with respect to the Hyades diagram.

3. In the same way add the diagram of the α Persei cluster. The dashed lines separate the cluster stars from a group of stars which have not the same proper motion. Actually a correction of $0^m.08$ for reddening and of $0^m.24$ for absorption should be applied, but this is almost negligible on our scale.

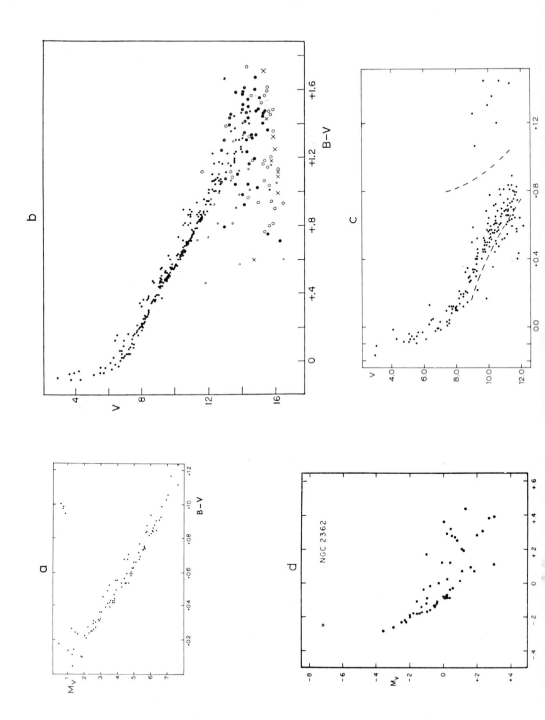

(4.) Add the diagram for NGC 2362 and perhaps for other open clusters. In the case of NGC 2362 the corrections for reddening and absorption have been applied. On the diagram here reproduced absolute magnitudes are given. You may use this diagram for the establishment of the ZAMS and to check the absolute magnitudes. Compare carefully whether your results agree with the scale on the figure.

5. You have now determined the ZAMS over a great part of its extent. Compare with the literature! (Sandage, A.: 1962, *Astrophys. J.* **135**, 351).

Moreover you have found signs of evolution in the clusters. The point where a cluster diagram branches off from the main sequence is characteristic for the age of the cluster. A star which is just leaving the main sequence must have changed its model appreciably and thus must have transformed about 13%, say, of its hydrogen into helium. This will have happened t years after the birth of the star. From the table below you are now able to determine the approximate age of the 4 clusters which we have studied.

6. We have at the same time found an excellent method in order to determine distances. The absolute magnitudes of the Hyades are very precisely known (see figure). The successive shifts of our diagrams correspond to an increase in the distance modulus and to an increase in the distance itself. Determine the scale of absolute visual magnitudes for the several clusters.

7. Adopting as a mean value for the Hyades a distance modulus of 3.0 and a distance of 40 pc, compute the distances of the other clusters.

Developing time t for main sequence stars

O7.5	2×10^6 years
B0	1×10^7
B5	5×10^7
A0	2×10^8
A5	6×10^8
F0	1.5×10^9
F5	3.5×10^9
G0	9×10^9

References

BAADE, W.: 1963, *Evolution of Stars and Galaxies*, chapters 9 and 10.
BLAAUW, A.: 1963, in *Stars and Stellar Systems*, Vol. III, 406.
SANDAGE, A.: 1968, in *Galaxies and the Universe*, Columbia University Press.
SAWYER HOGG, H.: 1959, in *Handbuch der Physik* Vol. 53, 132.
SEARS, R. L. and BROWNLEE, R. R.: 1965, in *Stars and Stellar Systems*, Vol. VIII, p. 619.

Fig. 71. Colour-Magnitude diagrams of open clusters. (a) The Hyades. After HECKMANN, O. and JOHNSON, H. L.: 1956, *Astroph. J.* **124**, 477. (b) The Pleiades. After JOHNSON, H. L. and MITCHELL, R. I.: 1958, *Astroph. J.* **128**, 31. (c) Cluster near α Persei. After MITCHELL, R. I.: 1960, *Astroph. J.* **132**, 68 .(d) NGC 2362. After JOHNSON, H. L. and HILTNER, W. A.: 1956, *Astroph. J.* **123**, 269. In the diagrams (a) and (d), absolute magnitudes are plotted; in (b) and (c), apparent magnitudes.

Preparation

For each student: rectangular coordinate paper.

Suitable clusters on which good photometric measurements have been made:

Hyades – O. HECKMANN and H. L. JOHNSON: 1956, *Astroph. J.* **124**, 477.

Pleiades – H. L. JOHNSON and R. I. MITCHELL: 1958, *Astroph. J.* **128**, 31.

Cluster near α Per – R. I. MITCHELL: 1960, *Astroph. J.* **132**, 68.

Praesepe – H. L. JOHNSON: 1952, *Astroph. J.* **116**, 640.

NGC 2362 – H. L. JOHNSON and HILTNER: 1956, *Astroph. J.* **123**, 269.

h and χ Persei – Ibidem.

Coma Berenices – H. L. JOHNSON and KNUCKLES: 1955, *Astroph. J.* **122**, 214.

Most practical are the diagrams, all on the same scale, published by K. A. BAR-KHATOVA and V. V. SYROVOY: 1963, *Atlas of Colour-Magnitude Diagrams for Clusters and Associations*, Moscow. See also: E. E. MENDOZA: 1967, *Boletin Tonanzintla*, **4**, 149.

Demonstration

G. ALTER and J. RUPRECHT: 1963, *The System of Open Star Clusters and our Galaxy. – Atlas*. Praha, New York, London.

G. ALTER *et al.*: 1958, *Catalogue of Star Clusters and Associations*, Praha.

MARKARYAN, B. E.: 1952, *Atlas Otkrytykh Skopleniy*, Moscow.

B34. THE DISTRIBUTION OF THE STARS
IN A GLOBULAR CLUSTER (L)

The Problem

You have received the photograph of a globular cluster. You notice how the stars, in this projection on a plane, are clustering in much closer packing near the centre, while the outer parts are much looser. The distribution law, describing this clustering, is of course of the highest interest for the understanding of the dynamics of this system and of its origin. However, the distribution *in space* must be a function of the radius r, different from the distribution *in projection on a plane*. We want to find this distribution in space.

Procedure

1. Divide the photograph in equally wide, horizontal strips, approximately symmetrical with respect to the centre, by putting light pencil marks along the frame of the photograph. We take the height of these strips as our unit of length. In this same unit we also measure the mean vertical distance of each strip to the centre. When speaking about the star density, we mean the number of stars within (this unit length)^3.

2. Now we count the number of stars F in the successive horizontal strips. In order to save time, you may consider only the even numbered strips. In one or two of the central strips the counts will not be reliable, they should be rejected. Moreover it is of advantage to average the numbers, referring to symmetrically located strips.

3. Plot the function $F(r)$, which describes how the number of stars varies as a function of r, the distance from the centre C to the strip.

4. From Figure 72 we see that

$$F(r) = \int_{p=0}^{p=R} 2\pi p \times f(s)\, \mathrm{d}p$$

Since $s^2 = r^2 + p^2$, we have: $p \times \mathrm{d}p = s \times \mathrm{d}s$; and:

$$F(r) = \int_{r}^{\sqrt{R^2 + r^2}} 2\pi s \times f(s)\, \mathrm{d}s$$

where R, the outer limit of the cluster, may be assumed to be very big, so that $\sqrt{R^2 + r^2} \simeq R$.

The derivative of this definite integral with respect to r is, according to the classical rule:

$$\frac{\mathrm{d}F}{\mathrm{d}r} = 2\pi R \times f(R) - 2\pi r \times f(r) \simeq -2\pi r \times f(r)$$

(assuming that $R \times f(R)$ approaches 0 for $R \to \infty$).
Consequently:

$$f(r) = -\frac{1}{2\pi}\frac{\mathrm{d}F}{\mathrm{d}r}.$$

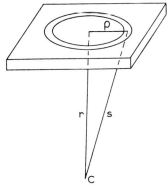

Fig. 72.

5. Find graphically the derivative dF/dr and calculate

$$f(r) = -\frac{1}{2\pi r}\frac{dF}{dr},$$

which represents the space density of the stars. Remember to indicate the units!

6. Plot $f(r)$ on log-log paper, and find the power of r which describes approximately the outward decrease. If you have no suitable paper of this kind, plot $\log f$ against $\log r$ on ordinary millimetre paper.

7. Find the absolute value of the units and determine the number of stars per cubic parsec.

(8.) Do the numbers $F(r)$ seem to approach a definite limit near an outer boundary? How far does the cluster actually extend?

References

CAMM, G. L.: 1952, *Monthly Notices Roy. Astron. Soc.* **112**, 169.
HOGG, H. S.: 1959, 'Star Clusters', in *Handbuch der Physik* **53**, 166.
SHAPLEY, H.: 1930, *Star Clusters*, Cambridge.
TEN BRUGGENCATE, P.: 1927, *Sternhaufen*, Berlin.
VAN BUEREN, H. G.: 1952, *Bull. Astron. Inst. Neth.* **11**, 403.

Preparation

For each pair: a photograph of a suitable cluster; square coordinate paper, log-log coordinate paper ($\log y$ against $\log x$).

Photographs of globular clusters are found, e.g., in the following publications: O. STRUVE: 1959, *Elementary Astronomy*, New York and Oxford, p. 275; *Public. Yerkes Obs.* **2**, plate 19; *Public. Lick Obs.* **8**, plate 54; *Contrib. Mt. Wilson Obs.* No. 177, plate 7; I. ROBERTS: 1899, *Photographs of Stars* etc., Vol. 2, London, plate 9; J. GAUZIT: 1960, *Images du Ciel*, Paris, plates 39 and 40; F. HOYLE: 1962, *Astronomy*, London, p. 264; A. EISENHUTH: 1967, *Das Weltall im Bild*, Graz.

B35. CLUSTER VARIABLES IN A GLOBULAR CLUSTER (L)

The Problem

Records of the globular cluster ω Cen B have been obtained at Johannesburg and later discussed by W. Chr. Martin (1937, *Ann. Sterrew. Leiden*). You will work on contact prints of these plates and derive the light curve of a variable star in this cluster.

Always turn the gelatine side of the plate away from you, avoid making scratches! The region to be studied is marked by a red circle.

First evening: to estimate the magnitudes

1. Orient the plate in the same sense as the photograph on paper. Identify the variable (Martin's Var. 8) and the 4 comparison stars a, b, c, d (Figure 73). If necessary, compare with Martin, p. 166, plate I.

2. Estimate the magnitude of the variable v with respect to the comparison stars, according to the notation a3v2b. Note the plate number, which is written with ink on the edge of the plate. Make similar estimates on the other plates.

3. When all 18 plates have been inspected, interpolate the magnitudes of the variable between those of the comparison stars. For each day when observations have been made, look up the corresponding Julian day. The magnitudes of the comparison stars are:

$$a = 14^m.09$$
$$b = 14^m.55$$
$$c = 14^m.80$$
$$d = 15^m.35$$

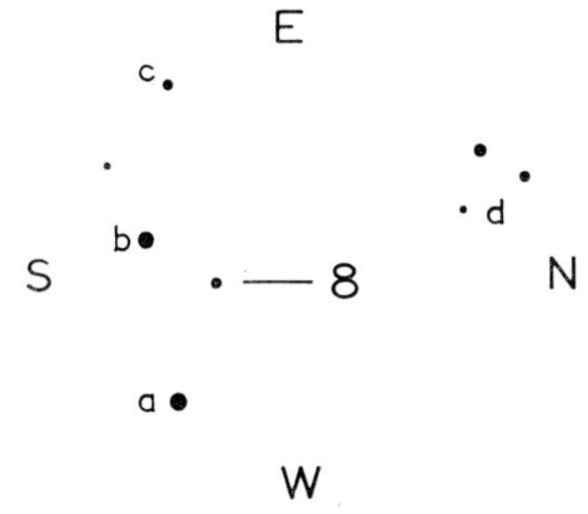

E

c.

·

b•

· d

S   · — 8   N

a •

W

Fig. 73.

4. Compare your results with those obtained by an experienced observer from the original plates and by means of a Schilt photometer.

Second evening: the determination of the period

5. We shall here make use of the magnitudes estimated by Dr. Martin. Plot the magnitudes against the date of observation ($1^d = 0.5$ cm, $1^m = 2$ cm). The upward direction should correspond with increasing brightness, i.e. decreasing magnitude. The observations give small sections of the light curve, some of them including the moments of quickly increasing brightness. First note for each of these observations whether the observations correspond to the ascending or to the descending part of the light curve; mark this on your drawing. Time intervals can be measured in a reliable way only from one ascending section to another or from one descending section to another.

6. Compare the observations on J.D. 37 and 38. Plot them on a scale, 10 times enlarged ($1^d = 10$ cm; $1^m = 2$ cm). Measure as precisely as possible the distance 37–38; it is found to be about 1 day. The period might thus be 1 day, but it could also be $0^d.5$. A decision is difficult because, due to local circumstances, the observations could be made only at intervals of about one day.

7. Compare now J.D. 41 and 52 (use only the observation at $41^d.466$, for which the brightness is intermediate between the points of J.D. 52). Plot again these observations on an enlarged scale, but shift the two sections towards each other by 10 days; by measuring the small, remaining distance, the total interval 41–52 is again better determined.

8. We have now to choose between a period near 1 day or another near 0.5 day. Divide the three intervals 37–38, 41–52 and 28–38 by suitable whole numbers and compare the three values which you find for the period. That assumption, according to which the three periods found present the smallest differences, is to be preferred. A still better decision may be obtained by using also the observations at 1495; at the same time the period will be very precisely determined. Ask the instructor to check the period obtained by you.

9. Now combine all observations so that they fit those of J.D. 52, by adding to the abscissae or by subtracting from them a whole number of periods. Plot your points on a large scale, e.g. $0^d.5 = 10$ cm and $1^m = 10$ cm. First transfer the point J.D. 41.466: if it does not fit accurately the ascending part of the light curve, the period should be very slightly modified. With this modified value all other points are then transferred.

10. From this value of the period it follows that your star is a typical cluster variable and that the absolute magnitude $\bar{M}_{pg}$ is about -0.2. Compare with the apparent magnitude $\bar{m}$ and determine the distance of ω Cen in parsec.

(11.) The coordinates of ω Cen are: $\alpha = 13^h 20^m.8$; $\delta = -46° 47'$. Look up this cluster in the Franklin-Adams charts and estimate its diameter: first in mm, then in radians, finally in parsec.

References

MARTIN, W.: 1937, *Ann. Leiden Obs.*, **17**. In this paper a much greater number of measurements of our variable are found (table 15). See further the literature, quoted at the end of exercise B34.

TABLE

Brightness of a Cluster Variable in ω Cen
(After MARTIN: 1937, *Ann. Leiden Obs.* (**17,** variable nr. 8.)

Jul. Day	m	Jul. Day	m	Jul. Day	m
+ 2426400					
28.4480	13.86	41.4168	15.35	53.4242	15.25
4771	14.09	4239	15.20	4314	15.01
4989	14.23	4452	15.02	59.3639	14.81
35.4295	15.04	4523	14.69	3857	14.94
4513	15.18	4660	14.14	4009	15.01
4731	15.27	52.3849	15.05	4151	15.09
37.4235	14.60	3994	14.63	60.3612	14.64
4380	14.76	4140	14.08	3916	14.81
4599	14.87	4285	13.89	4138	14.97
38.4287	14.38	4357	13.94	63.3674	13.97
4498	14.51	53.3878	15.33	3887	13.90
4716	14.63	3951	15.36	4031	14.15
				1495.4535	14.62 descend.
				1506.4093	14.68 descend.

Preparation

A set of contact copies of 18 original plates of ω Cen, to circulate among the students. The Franklin-Adams chart of this region.

For each pair: pulpit, magnifier.

Note. – An adequate set of 8 photographs, on which the variations of 6 stars in M 15 may be studied, have been prepared by O. Gingerich. To be purchased from: Sky Publishing Corporation, Cambridge, Mass.

B36. THE APEX OF THE SOLAR MOTION (L)

A. From Proper Motions

Because the sun (together with the earth) is in motion in the direction of the apex, all the stars appear to move from that apex and towards the antapex. However, this effect combines with the peculiar stellar motions, which are almost random.

Consider an area of the sky, located on the same meridian as the apex (Figure 74).

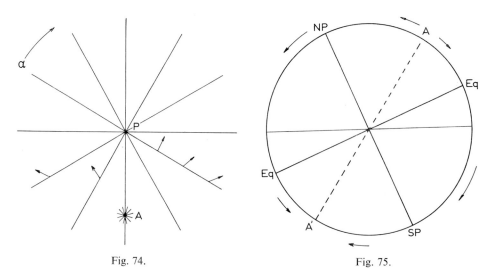

Fig. 74.   Fig. 75.

Among the stars in that area, there will be about as many for which the right ascension increases as for which it decreases. In areas towards one side of the meridian, there will be more stars with increasing r.a.; towards the other side the decrease of r.a. will prevail. Consequently, statistics of the component μ_α of the proper motion must allow a determination of the meridian of the apex.

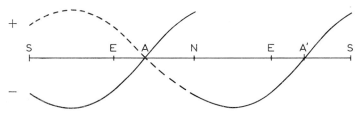

Fig. 76.   Value of $\bar{\mu}_\delta$ along the apex meridian.

Once this meridian is found, statistics of the μ_δ components along this meridian must yield the declination of the apex. Looking at the diagram (Figures 75, 76), we see that for stars moving away from the apex the sign of μ_δ is suddenly reversed when we pass the poles N or S; this is avoided if we give the opposite sign to all values of μ_δ on the arc NAS, while those on the arc NA'S keep their sign.

PROCEDURE

The Right Ascension of the Apex

1. Take a catalogue of proper motions (Boss, Schlesinger). We take groups of 50 stars, as near as possible to the meridians of 3^h, 6^h, 9^h, etc. Count how many of them have a positive component μ_α and how many have a negative component (P and N). We do not pay attention to the *value* of μ_α, only to the *sign*.

2. Plot the relative excess $(P-N)/(P+N)$ as a function of the right ascension. The zero points of this function correspond to the apex and to the antapex; find out which is which! If these two points are not distant by 12 hours, take means.

From the combined numbers P, N of all students a general curve may be derived, which gives even better results.

The Declination of the Apex

3. Now take the stars within a strip on both sides of the meridian which passes through the apex and the antapex. Make 12 groups according to declination, each group extending over $30°$. In each group count the positive and the negative values of μ_δ. One student of each pair reads the declinations of all stars having $\mu_\delta > 0$, the other puts a mark for each star in the corresponding group. The same for $\mu_\delta < 0$. (Near the poles any values of α are acceptable.)

4. Draw the fraction $(P-N)/(P+N)$ as a function of δ. Remember our remark concerning the reversal of sign μ_δ near the poles; change the sign over one half of the meridian.

This function has also two zero points, which should be distant by $180°$.

5. You have now right ascension and declination of the apex. Look up a *Star Atlas* and see in which constellation it is located. Compare with the literature. It will be seen that the result depends on the kind of stars with respect to which the determination has been made.

B. From Radial Velocities

In order to find the velocity of the sun through space, we shall determine its three components (Figure 77):

z, z' in the direction of the celestial poles ($\delta = 90°$, $\delta = -90°$)

x, x' in the plane of the equator, for $\alpha = 0$, $\alpha = 12^h$

y, y' in the plane of the equator, for $\alpha = 6^h$, $\alpha = 18^h$.

1. Take a catalogue of radial velocities (Campbell, Moore, Wilson). Look at the way it is arranged and find the meaning of the successive columns.

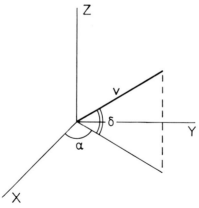

Fig. 77.

2. For 6 groups of stars, near the direction of the coordinate axes, as chosen above, determine the mean value of the radial velocities. In each group there should be at least 25 stars; differences of 20° or 30° with respect to the axes are still accepta-ble. Call these mean values: x, x', y, y', z, z'.

3. The mean values of x and x' should be equal but of opposite sign. The differ-ence of their absolute values gives an estimate of the statistical fluctuations or might be an indication for the K-effect (See exercise B37). Take $\bar{x}=(x-x')/2$ and similarly $\bar{y}$ and $\bar{z}$.

4. Finally, compute $v=\sqrt{x^2+y^2+z^2}$. Obviously your result, derived from such small material, has only a restricted value, but it gives the order of magnitude. It could not be derived from proper motions.

Derive the region of space towards which the sun moves, by computing α and δ of the apex (Figure 77). Avoid errors of sign!

5. Look up a *Star Atlas* and see to which constellation this direction of motion corresponds. Compare your result with that derived from proper motions.

TABULATION

Determination of α

	0^h	3^h	6^h	9^h	12^h	15^h	18^h	21^h	24^h
P									
N									
$P-N$									
$P+N$									
$P-N$									
$P+N$									

Determination of δ

	90°	60°	30°	0°	30°	60°	90°	60°	30°	0°	30°	60°	90°
P													
N													
$P-N$													
$P+N$													
$\dfrac{P-N}{P+N}$													

$$\alpha = 18^h \qquad\qquad\qquad \alpha = 6^h$$

References

This exercise is inspired by the text of RUSSELL, DUGAN, and STEWART: 1938, *Astronomy* II, pp. 657–661.
See also:
BOK, B.: 1965, *Seminars on the Structure and Dynamics of the Galaxy*, Mt. Stromlo Observatory, ch. V.
DELHAYE, J.: 1965, in *Stars and Stellar Systems*, Vol. 5, 61.
EDDINGTON, A. S.: 1914, *Stellar Movements*, Chapter V.
LANDOLT-BÖRNSTEIN: 1965, *Astronomie und Astrophysik*, **8**, 3, pp. 627.
VON DER PAHLEN, E.: 1937, *Lehrbuch der Stellarstatistik*, Leipzig, p. 691–715.

Preparation

For each group: a catalogue of proper motions; a catalogue of radial velocities; *Star Atlas*.
List of catalogues in: LANDOLT-BÖRNSTEIN 1965, VI/1; p. 274.

Proper motions and radial velocities

GLIESE, W.: 1957, *Mitt. Astr. Rechen-Instit. Heidelberg*, Ser. A, Nr. 8.
Yale Catalogue of Bright Stars, 1964, three editions, New Haven.

Proper Motions

BOSS, B.: 1937, *General Catalogue*, Vol. 2–5, Washington.
BOSS, L.: 1910, *Preliminary General Catalogue*, Washington.
Connaissance des Temps, 1914.
Star Catalogue, 1966, Smithson. Astroph. Obs.

Radial Velocities

CAMPBELL, W. W. and MOORE, J. H.: 1928, *Public. Lick Obs.* **16**, 347.
MOORE, J. H.: 1932, *Public. Lick Obs.* **18**.
VOÛTE, J.: 1928, *Ann. Observ. Lembang*, **3**.
WILSON, R. E.: 1953, *General Catalogue of Radial Velocities*, Washington.

Note: in order to reduce by half the number of copies necessary, half of the students may work on proper motions, while the other half is studying radial velocities; after which they exchange roles.

B37. PREFERENTIAL DIRECTIONS IN
THE STELLAR MOTIONS (L)

In the preceding chapter we have assumed that the motions of the stars are random. More extensive investigations have shown that the peculiar motions show an anisotropy: there is a preference for the direction near $\alpha = 18^h$, $\delta = -12°$ and for the opposite direction (*vertices*). Let u, v, w be the velocity components of which u corresponds with the vertex; then the number of stars with components between u and $u + du$, v and $v + dv$, and $w + dw$ is found to correspond approximately to the function:

$$N(u, v, w)\, du\, dv\, dw = c \times e^{-h^2 u^2 - k^2(v^2 + w^2)}\, du\, dv\, dw.$$

Surfaces of equal frequency ($N = $const) are ellipsoids.

We intend to verify that formula and to determine the axes of the ellipsoid. (The computations for the u-component and those for the v-component can be made by different groups of students, who later combine their results.)

1. Study a catalogue of radial velocities and find out how it is arranged and what is the meaning of the different columns.

Take at random 200 stars, 100 near the vertex and 100 in the opposite direction; suitable boundaries for these areas are:

$$\alpha = 17^h \text{ to } 19^h, \qquad \delta = -25° \text{ to } 0°$$
$$\alpha = 5^h \text{ to } 7^h, \qquad \delta = 0° \text{ to } +25°.$$

If necessary, these boundaries may be somewhat enlarged. Stars of very early type are not counted.

Count the number of stars, having radial velocities s within each of the following boundaries: (km/sec)

0 to 10	10 to 20	20 to 30	30 to 40	40 to 50	50 to 60
0 to -10	-10 to -20	-20 to -30	-30 to -40	-40 to -50	-50 to -60

It is most instructive, to watch how such a curve builds up gradually, becoming smoother and more regular. Make a plot of the distribution function $N(s)$.

2. This function can be approximated by a Gauss curve. This would be in agreement with the statement in the introduction, since we have considered only the u- components and should therefore expect: $N(u)\, du = c' \times e^{-h^2 u^2}\, du$.

That the curve is not symmetric with respect to the zero-point is due to the parallactic component of the sun's motion.

3. *To fit a Gauss-curve to an empirical distribution*
A. Find the mean value of the abscissae: $\bar{s} = \Sigma ns / \Sigma n$, then make a new scale with abscissae $u = s - \bar{s}$.

B. Compute $\overline{u^2} = \Sigma nu^2/\Sigma n$, which we identify with $1/2h^2$.

In both computations the summations are not made star by star but group by group, which works much more quickly and is a very good approximation. For the determination of $\bar{u}^2$ we combine the stars of both vertices, taking, of course into account the sign of the radial velocities; the same applies to the drawing of the Gauss curve. This curve is now:

$$N(u) = \frac{Mh}{\sqrt{\pi}} e^{-h^2u^2} = \frac{Mh}{\sqrt{\pi}} exp\left(-\frac{u^2}{2\overline{u^2}} \right),$$

where M is the total number of stars $= \int N(u)\, du$. Draw this curve and compare with the observed curve. Are there any 'high-velocity stars'?

4. Repeat the whole operation for the motions towards a pair of points at $90°$ from the vertices, for example near $\alpha = 18^h$, $\delta = +80°$ and the opposite direction. (Look on a star-globe.) We find now the curve $N(v) = c'' \times e^{-k^2v^2}$.

Compare the dispersion $1/k$ with that of the u-curve $1/h$. *This difference in dispersion is the fundamental phenomenon* formerly called '*star streaming*', which has been explained later as a consequence of the orbital motion around the galactic centre, which is elliptic, especially for stars of spectral types dK and dM.

5. Compare the values which you have derived for the axes of the velocity ellipsoid with those found in the literature from much more extensive material.

(6.) Are the values of $\bar{s}$ the same in absolute value for the two vertices? If not, it would seem as if there was, apart from the solar motion, a general expansion (or contraction) of the galaxy in our surroundings. What value do you get for this 'K-term'? This may be partly due to systematic errors in the radial velocities, partly to gravitation effects in massive stars, partly to dissipation of a local cloud of early stars, for which it is most apparent.

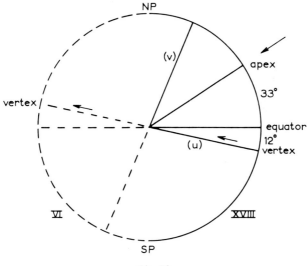

Fig. 78.

(7.) From the values of $\bar{s}_u$ and $\bar{s}_v$, corrected for the K-effect, you may derive the velocity of the sun towards the apex $(\alpha = 18^h, \delta = +30°)$. Since the two velocity vectors and the apex are almost in the plane of the hour circle of 18^h, we may write:

$$\text{velocity of the sun} = s_0 = \sqrt{\bar{s}_u^2 + \bar{s}_v^2} \text{ (Figure 78)}.$$

Moreover $\bar{s}_u = s_0 \cos \vartheta$ and $s_v = s_0 \sin \vartheta$, where ϑ is the angle between apex and vertex. Compare with the results of B36.

References

Bok, B.: 1965, *Seminars on the Structure and Dynamics of the Galaxy*, Mt. Stromlo Observatory, ch. VI.
Delhaye, J.: 1965, in *Stars and Stellar Systems*, Vol. 5, 61.
Landolt-Börnstein: 1965, *Astronomie und Astrophysik*, **8**/1, pp. 636–637.
Oort, J. H.: 1965, in *Stars and Stellar Systems*, Vol. V, 474.
Wilson, R. E.: 1941, *Astroph. J.* **93**, 511.
Woolley, R.: 1965, in *Stars and Stellar Systems*, Vol. V, 89.
The exercise is inspired by the following paper:
Williamson, R. E.: 1941, *Astroph. J.* **93**, 511.
For a quick reduction of galactic coordinates to equatorial coordinates, use the maps 17 and 18 in *Norton's Star Atlas*.

Preparation

For each pair: a catalogue of radial velocities; rectangular coordinate paper.
Catalogues of radial velocities are listed in exercise B36.
A star globe.

B38. THE MILKY WAY (S)

We intend to draw the course of the Milky Way between the stars. This observation has to be made on a clear night, in the field, at least 5 km from the outskirts of the city. The moon is disturbing and so are street lights even at considerable distances. Our pocket lamps must give very subdued light.

1S. THE OBSERVATION. When your eyes are adapted to the darkness, we follow the general course of the Milky Way and identify the principal constellations along it. You find this part of the sky on the star maps of Rohrbach (or on other atlases where the Milky Way is not drawn). Try to determine accurately the heart line of the Milky Way and to plot several of its points on the map. In doing so, we should keep our hand before the flash lamp and admit only the strict minimum of light. Note each of the points which you have chosen with a light pencil mark and then put out the light immediately; even when taking this precaution you will find that the eye takes a minute or so to recover its full sensitivity.

First study the Southern part of the Milky Way, going as far down to the horizon as possible. Then pass through the zenith towards the Northern side, and there also try to reach the horizon.

Now and then you will notice a bifurcation, as e.g. near Perseus. The longer you look, the more details will become visible, darker and brighter clouds. Always look for the 'centre of gravity' across the bright belt.

2L. WORKING-OUT YOUR NOTES. Copy your marks carefully on a map of the Northern sky which you have received. Draw a smooth line through them: this is approximately the galactic equator.

3L. Find the minimum distance of the pole to that circle and derive its inclination to the equator. Also determine the position of the nodes, where the galactic equator crosses the celestial equator.

Compare the results with those in the literature.

References

PANNEKOEK, A.: 1920, Die Nördliche Milchstrasse, *Ann. Sterrenw. Leiden*, **11**[3]. – 1933, *Publ. Amsterdam*, Nr. 3.
BOK, B. J. and BOK, P. F.: 1957, *The Milky Way*, Cambridge, Mass., ch. 1.

Preparation

For each group: flashlight (lamp underloaded, light strongly dimmed by red paper); star maps as used for meteor observation (e.g. ROHRBACH: 1907, *Sternkarten in gnomonischer Projektion*, Gotha); a general map of the sky, such as NORTON's *Northern Survey Map*, but without the drawing of the Milky Way. Or the star maps in *Sky and Telescope*, on which you can draw with white chalk (with some difficulty).

B39.   DARK NEBULAE (L)

The Problem

In many fields of the Milky Way the stars appear to be surprisingly rare. It may be shown that this is due to huge dust clouds, which locally weaken the light of the background. Information about such clouds may be obtained by star counts (a) in a normal field, (b) in an obscured field. Such counts have to be made for each of the magnitude classes separately.

As an object for study we use the photographic Franklin-Adams *Map of the Sky* or Vehrenberg's *Photographischer Stern Atlas. These very valuable documents must be carefully handled*, no marks may be made on them!

Procedure

1. First look up the region to be studied in the atlas of Khavtassi and in the photographs of Barnard or Ross-Calvert. Compare with the corresponding sheet of the Vehrenberg Map, choose an 'obscured region' and an 'unobscured region'.

Place a celluloid plate on the chart and trace on it the contours of the regions to be compared. – In doing so, avoid making scratches or putting marks of any kind on the chart!

2. In order to estimate the brightness of the stars, you will receive a scale of stellar images on a strip of paper, corresponding with a scale of increasing brightness. A few of them have been selected as typical, each one for a broad class of brightness. Try to remember roughly how these images look. Number them.

3. Now put somewhere on the obscured region a celluloid plate on which a rectangular frame has been traced (2 mm × 3 mm). Estimate the brightness number for every star which is seen within the frame, scanning systematically from left to right. Use a magnifier.

Make a provisional trial first in order to get some experience. Then definitely assign to each star within the frame in the obscured region a brightness number; these numbers are scored by your partner. Shift the frame till you have counted in all about 300 stars. Try to reach the faintest images as well. A sufficient number of the brightest stars will not easily be found; in that case, count these stars over a wider field.

4. Repeat the same observations for an unobscured field. Try to keep the scale of the estimated brightness constant by often comparing with your standard scale.

5. Plot the numbers found in each of the brightness classes; if necessary smooth the curves slightly. Then compute the ratio of the star numbers:

$$V = \frac{A' \text{ (obscured)}}{A \text{ (unobscured)}}$$

for each of the brightness classes.

6. In order to reduce your 'brightness numbers' roughly to the ordinary magnitude scale, compare your standard stellar images with the Vehrenberg (or Franklin-Adams) chart of the North-Pole Region, for which careful photometry has been made.

For the 'normal' number of stars $\bar{A}(m)$ in the unobscured region we prefer to take the numbers $\bar{A}(m)$ which have been found by counts over the whole sky at similar galactic latitudes. Plot $\log \bar{A}(m)$ as a function of m; then find from your observed $V(m)$ the values of $\log \bar{A}'(m)$, which you will plot in the same graph (*Wolf diagram*).

7. Discuss the two curves $\bar{A}(m)$ and $\bar{A}'(m)$. Is it possible to tell approximately around which magnitudes the dark nebula makes itself felt and begins to obscure the background appreciably? What is the approximate distance of the nebula? – What is the total obscuration suffered by the farthest stars?

Note. If stars of all magnitudes are about equally obscured, this means that the cloud is quite near to the sun. If the brighter stars are *more* numerous in the obscured region, this may be explained either by accidental deviations, the number of stars counted being too small; or by an organic relation with the cloud, which might have increased the rate of star formation; or by a neighbouring cloud, veiling the region which was considered to be 'unobscured'.

TABULATION

Unobscured field

brightness	1	2	3	4	5	....
....	....	....	....	....	....	....
....	....	....	....	....	....	....
Total	....	....	....	....	....	....

Obscured field

brightness	1	2	3	4	5	....
....	....	....	....	....	....	....
....	....	....	....	....	....	....
Total	....	....	....	....	....	....

$m=$	5	6	7	8	9	....
Ratio $V(m)$						
$\log V(m)$						
$\log \bar{A}(m)$						
$\log \bar{A}'(m)$						

TABLE

Values of $\log A(m)$

After SEARES, F. H., VAN RHIJN, P. J., JOYNER, M. C., and RICHMOND, M. L.: 1925,
Astrophys. J. **62**, p. 368.

m	Galactic latitude			
	$0°$	$30°$	$60°$	$90°$
4.0	8.219	7.898	7.734	7.680
4.0	8.447	8.127	7.963	7.910
5.0	8.674	8.355	8.191	8.139
5.5	8.900	8.582	8.417	8.364
6.0	9.125	8.807	8.641	8.584
6.5	9.349	9.030	8.863	8.802
7.0	9.572	9.252	9.082	9.016
7.5	9.794	9.471	9.299	9.227
8.0	0.016	9.688	9.512	9.433
8.5	0.236	9.901	9.721	9.634
9.0	0.455	0.113	9.924	9.830
9.5	0.673	0.323	0.121	0.022
10.0	0.889	0.529	0.312	0.209
10.5	1.102	0.731	0.498	0.390
11.0	1.314	0.928	0.679	0.565
11.5	1.525	1.120	0.854	0.735
12.0	1.733	1.308	1.024	0.898
12.5	1.938	1.491	1.189	1.054
13.0	2.141	1.668	1.348	1.205
13.5	2.339	1.839	1.502	1.351
14.0	2.532	2.003	1.649	1.491
14.5	2.720	2.160	1.789	1.625
15.0	2.902	2.310	1.922	1.752
15.5	3.078	2.453	2.048	1.873
16.0	3.248	2.589	2.167	1.988
16.5	3.412	2.716	2.279	2.095
17.0	3.571	2.836	2.384	2.104
17.5	3.725	2.950	2.482	2.287
18.0	3.874	3.058	2.573	2.374
18.5	4.017	3.159	2.657	2.456
19.0	4.153	3.253	2.734	2.531
19.5	4.282	3.339	2.803	2.597
20.0	4.405	3.417	2.864	2.654
20.5	4.522	3.487	2.917	2.702
21.0	4.631	3.547	2.961	2.743

References

BECKER, W.: 1950, *Sterne und Sternsysteme*, Vol. V, Dresden und Leipzig, p. 25.
BOK, B. J. and BOK, P. F.: 1957, *The Milky Way*, Cambridge, Mass., ch. 7.

Preparation

Carefully selected sheets from a photographic map of the sky. Either the *Franklin-Adams Chart*; or H. VEHRENBERG: 1963, *Photographischer Stern Atlas*, Düsseldorf, Treugesell-Verlag (from $\delta = -26°$ up to $\delta = 90°$).
Demonstrate:
BARNARD, E. E.: 1927, *A Photographic Atlas of the Milky Way*, Washington.
ROSS, F. E. and CALVERT, M. R.: 1934, *Atlas of the Northern Milky Way*, Chicago.
KHAVTASSI, D. Sh.: 1960, *Atlas of Dark Galactic Clouds*, Tbilisi.

For each pair: celluloid plate; celluloid plate with 2 mm × 3 mm frame; scale of star images obtained by making a series of photographic exposures by means of a telescope, shifting the plate each time over a few millimetres and multiplying the preceding exposure time by 1.5; this scale is copied on glass or film, then printed on paper (the star images are now black on a white background); square coordinate paper; magnifier.

The North Polar Sequence is difficult to identify, because usually the drawings show only those stars which have been selected for photometry! In the following publications, however, complete drawings are found, which makes the identification easy and pleasant:
KULIKOWSKY, P. G.: 1961, *Spravočnik*, p. 404; and especially ROUSSEAU, J. and SPITE, F.: 1958, *Contr. Labor. Astron. Lille*, No. 9, which has to be used in combination with *Ann. Harv. Obs.* **71**, Nr. 3, p. 49 or with *Astroph. J.* **41**, p. 223.

Dark Galactic Clouds
Suitable Objects for investigation

Khavtassi number	Constellation	α	δ	l^{I}	l^{II}	b	Literature
278 273 282	Tau	4^h 20^m	$+28°$	$138°$	$171°$	$-13°$	Z. Astroph. **6** (1933), 259
264	Aur	4 30	$+38$	132	165	-5	Z. Astroph. **17** (1939), 191
447	Crux	12 48	-62	270	303	0	Z. Astroph. **8** (1934), 66
556	Oph	16 23	-24	327	0	15	Z. Astroph. **3** (1931), 369
612	Oph	17 10	-22	$329\frac{1}{2}$	$2\frac{1}{2}$	$7\frac{1}{2}$	Z. Astroph. **4** (1932), 365
600	Oph	17 25	$-25\frac{1}{2}$	329	2	$2\frac{1}{2}$	Z. Astroph. **3** (1931), 261
744	Scu	18 50	-5	356	29	-3	Astron. Nachr. **229** (1926), 1

Other, well-investigated objects are listed by LYNDS, B. T.: 1968, in *Stars and Stellar Systems*, Chicago, Vol. VI, p. 131.

B40. THE SPIRAL STRUCTURE OF THE GALACTIC CLOUDS, DERIVED FROM 21 CM MEASUREMENTS (L)

The hydrogen clouds in our Galaxy emit a monochromatic radiation at a wavelength of 21 cm. Figure 79 reproduces spectral records, obtained by the Dwingeloo radio telescope, when directed successively to points along the galactic equator at intervals of 5° in galactic longitude.

It appears that the line often shifts towards the red or the violet side of the frequency proper. This we attribute to Doppler effects. The profile is in general complex: along the line of sight, several hydrogen clouds may be found, having different radial velocities with respect to the sun; they are transparent and we see their integrated radiation. From the scale under the diagram the relation between the wavelength shift and the relative radial velocity is obtained directly.

The objects in the Galaxy rotate about a common centre C in almost circular orbits; their angular velocity $\omega(R)$ decreases outward, as is also the case in the planetary system. The function $\omega(R)$, determined partly from observations, partly from theory is drawn in Figure 81.

We shall now derive the relation between the distance of an emitting object P and its radial velocity relative to the sun S. The velocity components along the line SP, subtracted, yield (Figure 80):

$$V_{rad} = R\omega \sin l - R_0\omega_0 \sin l^{II}.$$

But in triangle SCP we have: $R \sin l = R_0 \sin l^{II}$.

Consequently

$$V_{rad} = R_0 \{\omega(R) - \omega_0\} \sin l^{II}$$

The values adopted were $R_0 = 8.2$ kpc, $\omega_0 = 26.4$ km/sec·kpc in the original publication of Van de Hulst, Muller and Oort. At present a value $R_0 = 10$ kpc would seem preferable.

Procedure

1. Select the first (left) maximum of the profile at $l^I = 35°$. Calculate consecutively: $\Delta\lambda$, V_{rad}, $\sin l^{II}$, $\omega - \omega_0$, ω, then read R from Figure 81.

2. Plot the cloud in a polar diagram, representing the galactic plane. The sun is drawn 5 cm above the galactic centre C. From S draw the line of sight and find the position of the cloud.

Remember that the distance R is measured *from the galactic centre C*.

3. On the neighbouring records the corresponding maximum is recognized and may be interpreted similarly. Follow it towards smaller and towards larger galactic longitudes, so far as the continuity is entirely clear. In order to save time, you might take into account the profiles at intervals of 10° in longitude only.

4. Do the same with the other maxima, in so far as they are clearly indicated.

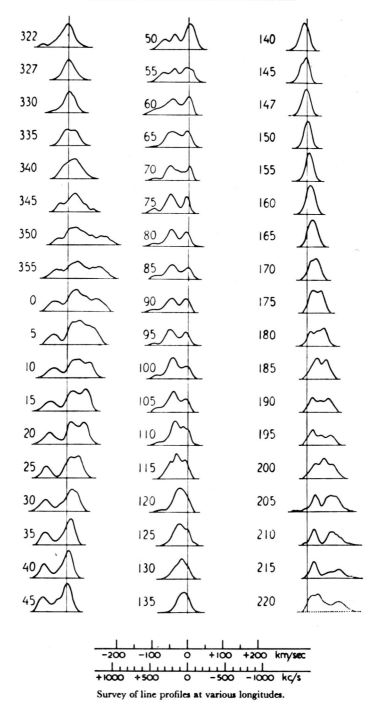

Survey of line profiles at various longitudes.

Fig. 79. Profiles of the hydrogen emission line at 21 cm. After VAN DE HULST, H. C., MULLER, C. A., and OORT, J. H.: 1954, *Bull. Astron. Inst. Neth.* **12**, 125. The galactic longitudes l^{I} given for each curve, should be reduced to the modern designation $l^{\mathrm{II}} = l^{\mathrm{I}} + 33°$, used in Figure 80.

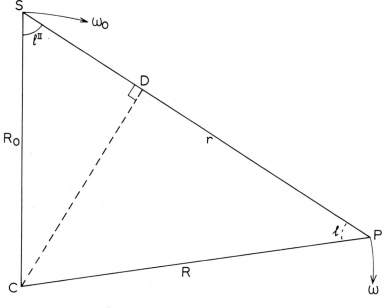

Fig. 80.

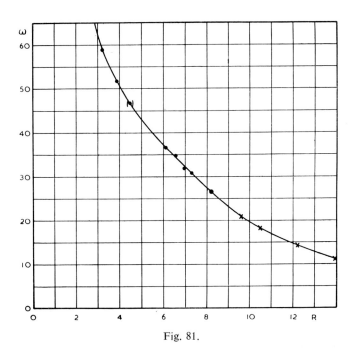

Fig. 81.

However, when R is found smaller than R_0, an ambiguity occurs: on the line of sight, *two* points have a distance R to the centre. This ambiguity cannot be solved without additional data and considerations. – We shall therefore limit our investigation to the region outside the solar orbit. This means:

for $0 < l^{II} < 180°$ use only the maxima for which $\Delta\lambda < 0$;
for $180° < l^{II} < 360°$ use only the maxima for which $\Delta\lambda > 0$.

5. Connect the plotted positions of the hydrogen clouds by smooth lines, in so far as continuity is suggested by the successive profiles. Distinguish the Orion arm, through the sun, and the Perseus arm, 2 kpc outside. The general spiral pattern appears clearly, the arms are fairly tightly wound and are almost circular in form. The *detailed* distribution however is still tentative.

TABULATION

l^{II}	$\Delta\lambda$	$V_{rad\ (km/sec)}$	$\sin l^{II}$	$R_0 \sin l^{II}$	$\omega - \omega_0$	ω (km/sec·kpc)	R (kpc)
....	....	....	....	....	....	....	....
....	....	....	....	....	....	....	....

References

Bok, B. J. and Bok, P. F.: 1957, *The Milky Way*, Cambridge, Mass., p. 242.
Kerr, F. J. and Westerhout, G.: 1965, in *Stars and Stellar Systems*, Vol. V, p. 167.
Van de Hulst, H. C., Muller, C. A., and Oort, J. H.: 1954, *Bull. Astron. Inst. Neth.* **12**, 117.

Preparation

For each pair: protractor; sine tables.

Recent data, adapted to $R_0 = 10$ kpc, in the survey by Schmidt, M.: 1965, *Stars and Stellar Systems,* Vol. V, p. 528.

INDEX OF SUBJECTS